SEX, SCIENCE AND PROFITS

Terence Kealey is a clinical biochemist and Vice-Chancellor of the University of Buckingham. He is author of *The Economic Laws of Scientific Research* and writes regularly for *The Times* and the *Sunday Telegraph*.

ALSO BY TERENCE KEALEY

The Economic Laws of Scientific Research

TERENCE KEALEY

Sex, Science and Profits

How People Evolved to Make Money

VINTAGE BOOKS
London

Published by Vintage 2009

2 4 6 8 10 9 7 5 3 1

First published in Great Britain in 2008 by
William Heinemann

Vintage
Random House, 20 Vauxhall Bridge Road,
London SW1V 2SA

www.vintage-books.co.uk

Addresses for companies within The Random House Group Limited
can be found at: www.randomhouse.co.uk/offices.htm

The Random House Group Limited Reg. No. 954009

A CIP catalogue record for this book
is available from the British Library

ISBN 9780099281931

Penguin Random House is committed to a sustainable future for
our business, our readers and our planet. This book is made from
Forest Stewardship Council® certified paper.

Printed and bound in Great Britain by Clays Ltd, St Ives plc

For Felicity Bryan, my agent, without
whom this book would not have been finished

Contents

Prologue

I
The Prophet of Progress

'Knowledge is power.' Francis Bacon, *Religious Meditatations, Of Heresies*, 1597

The concept of progress is surprisingly recent. The idea is only 400 years old. It was formulated in 1605 by an Englishman, Sir Francis Bacon (1561–1626), in his book *The Advancement of Learning*. Before Bacon, people had supposed history to be circular, not directional, and both Plato and Aristotle had asserted that history was cyclical, with republics and kingdoms replacing each other to no outcome. As Ecclesiastes 1:9 says, 'there is no new thing under the sun.'

In his 1920 book *The Idea of Progress* J. B. Bury, the Cambridge historian, noted that an ancient expression of improvement had in fact been made in Daniel 12:4: – 'many shall run to and fro, and knowledge shall be increased' – but that verse was overlooked for millennia by people who did not understand it: how could knowledge be increased when it had already been revealed by God? Democritus had hinted at the notion of progress during the fourth century BC, as had the Oxford scientist Roger Bacon during the fourteenth century AD, but only with Sir Francis Bacon was the concept described. He called it 'progression', the addition of new knowledge to old.

Bacon came to the idea of progress in 1588, when the Spanish launched the Armada against England. Though England, helped by the weather and the Dutch, beat off the Armada, Bacon never forgot

3

the fear. Indeed, Bacon's future secretary, the philosopher Thomas Hobbes, was born prematurely after his mother was shocked into labour on learning of the sighting of the Armada. And Mrs Hobbes was right to have been shocked: by 1588 the king of Spain was a man of vast power, who not only ruled significant parts of Europe (Portugal, the Low Countries, much of Italy, parts of Burgundy and so on) but who also straddled the globe, from Macau and the Philippines in the east to Mexico and Peru in the west. Indeed, he so girdled the world that the Philippines in the east were administered from Mexico in the west. He might easily have added England to Spain's list of conquests.

Bacon wanted to know how Spain had become so powerful. So he read the histories, including Gomes Eannes de Azurara's *The Chronicle of the Discovery and Conquest of Guinea* (1450), João de Barros's *Décadas da Ásia* (1552–63) and Antonio Galvano's *Summary of the Discoveries of the World* (1555), and he concluded that Spain had created its empire by technology: 'and like as the West Indies had never been discovered if the use of the mariner's needle had not been first discovered.' In 1620 Bacon wrote: 'Printing, gunpowder and the magnet [compass] . . . have changed the whole face and state of things throughout the world.' In his most famous quote, he said: 'Knowledge is power.'

But who had created Spain's technology, knowledge and power? From his readings of the chroniclers Bacon concluded that the hero of the story was not a Spaniard but a Portuguese, Prince Henry the Navigator. It was Henry who had launched the modern world of progress. How?

II
For God and Profit

'He that increaseth knowledge increaseth sorrow.' Ecclesiastes 1:18

There are two major reasons for doing science: to discover something new for its own sake or to discover something new for the sake of making money. On either score Henry the Navigator (1394–1460) can claim to have been one of the world's greatest scientists. Before him, people subscribed to a false cosmography and Portugal, his country, was poor. By the time his work was complete the Europeans had transformed world geography and Portugal was one of the richest countries on earth.

Henry was the third son of João (John) I, king of Portugal, and of Philippa, his queen. Philippa was the daughter of John of Gaunt, so Henry was half-Portuguese and half-English. But Portugal, when Henry was born, was a precarious little nation. For hundreds of years it had been ruled by Muslims from North Africa, and they had been expelled by the native Christians only the century before Henry was born.

Portugal was poor because it was excluded from the centres of international trade. The rich states of Europe were then the trading city-states, like Bruges, and *the* wealthy states were the ports of the Italian peninsula, including Venice and Genoa, which traded the spices, silks and gems of India, China and the islands of the East. Those trades were as old as European civilization itself: 'and, behold, a company of Ishmealites [Arabs] came from Gilead with their camels bearing spicery and balm and myrrh, going to carry it down to Egypt'. (Genesis 37: 25).

The spices would be re-exported from Egypt to Europe, whose people have always salivated over the peppers and gingers and cinnamons and saffrons and asafoetida and cardamoms and cloves of Persia, India, Malaysia and Indonesia (pepper and ginger are Indian loan words). The spice trade was, therefore, one of the most lucrative trades of the day, and the people who controlled it grew rich. Yet those lucky people were not the consumers in Europe, nor the producers in Africa and Asia, but the intermediaries, who were primarily the Muslim merchants of North Africa and the Middle East.

Strategically placed, the Arabs and Ottoman Turks allowed the merchants of the Italian city states to penetrate as far as the great bazaars of Damascus and Alexandria but no further. And, being monopolists, the Muslims charged monopoly prices. When Vasco da Gama was eventually to reach India in May 1498 he was to find the local price of pepper to be only 3 ducats per quintal, whereas in Venice it was twenty-seven times higher at 80 ducats. Nutmeg was even more valuable to shippers: when the Dutch reached the Banda Islands in the Indian Ocean, they found its local price to be one-six hundredth of its price in London or Paris. And, in 1563, 100 pounds of Ceylon cinnamon was in Europe worth 10 pounds of gold, while rhubarb, then a rare exotic medicine, was worth its actual weight in gold.

For as long as the Arabs and other middlemen dominated the trade, their inflated prices were to ruin their European customers, but the spice trade had always ruined its customers. As long ago as the first century AD Pliny was complaining that the traffic in pepper was depleting Rome of its gold reserves. The Roman satirist Persius wrote:

> Our merchants, money-making
> Sail to India, which is baking
> In exchange for Roman ware
> Pepper and medicines back they bear

Enraged by the ruinous prices, the Romans had eventually smashed the monopolies, sending first their armed galleys and then their merchant ships down the Red Sea to India (which is how Doubting Thomas got to be buried in Madras, Chennai). And inspired by the Roman example, Henry the Navigator decided that he, too, would break the monopolies of his own time – not by sailing via the Mediterrannean and Red Seas (which were closed to him by the Italians, Arabs and Turks) but via Ceuta, the Arab port in North Africa.

Ceuta was a logical target for Portuguese aggression. Possessing no fewer than 24,000 shops that overflowed with spices, rugs, gold, slaves, silver and jewels, it was tantalizingly rich. It was also tantalizingly close, lying on the northern coast of what is now Morocco, opposite Gibraltar and just a few hours from Portugal by ship. The Portuguese conquest of Ceuta would, moreover, gratify a maritime nation whose soldiers had for centuries been killing Muslims and who would not mind reaching out to kill a few more.

To ensure the success of their attack, the Portuguese took every precaution. First, they got the pope to promise immediate entry to Heaven to any crusader who died fighting the infidels. Second, Prince Henry built a fleet in Oporto. And third, King João and Prince Henry awaited the appropriate omens. These, fortunately, fell thick and fast: a monk near Oporto witnessed the Virgin Mary handing King João a glittering sword, which confirmed that God wanted the Muslims to be slaughtered; there was an eclipse of the sun, which could only guarantee a Portuguese victory; and, best of all, Queen Philippa fasted so dramatically that, in her emaciated state, she died of the plague, which was obviously miraculous because, on her deathbed, she handed her family pieces of the True Cross, and she was actually blessing the expedition to Ceuta as she expired.

So on 24 August 1415 the Portuguese assaulted Ceuta and with such vigour that, within a few hours, it fell. Only eight Portuguese were killed in the assault, but during the subsequent sacking of the town the numbers of Muslim dead soon reached thousands. European Christendom had broken out of Europe, to stride over the sands of North Africa, and this time – unlike the failed Crusades of a few centuries earlier – there would be no slinking back.

Yet after Portugal's victory Ceuta's economy collapsed. Its wealth had been built on Muslim trade, and no follower of Muhammad was going to trade with the murderous Christian invaders. Indeed, Ceuta soon became a liability to the Portuguese, who had to maintain an expensive garrison against counterattack. But the governor of that garrison was Prince Henry, and he soon discovered that, although Ceuta's trade had been undertaken by Muslims, they were only the middlemen – as usual.

Ceuta was not a spice-trading city but a gold-trading one. The gold came from equatorial Africa, from the tropical rainforests of the Senegal River basin, whose tribesmen extracted it by open-cast mining. The Arabs transported it from Senegal to the Mediterrannean. Their caravans would embark from Ceuta and cross the Atlas Mountains and the deserts of the western Sahara to reach, after some twenty days, the Senegal River. There they bartered salt, beads and other artefacts for gold before returning with their booty.

Henry wanted to capture the Senegal gold trade, but the overland route was blocked by the armies of the Mali Empire, then so impenetrable that people wove myths about Timbuktu and its other cities. So Henry opted to sail there. Unfortunately, however, in 1415 the Atlantic was still the Sea of Darkness, almost unexplored. During the previous centuries the Genoese had apparently discovered the Azores, Madeira and the Canaries, and in 1291 the Vivaldi brothers from Genoa had even tried to sail round Africa, but the discoveries had not been systematically chronicled, the details had been lost, and the unfortunate Vivaldis were never seen again. The Atlantic was the Sea of Doom.

Yet, as governor of Ceuta, Henry commanded a small fleet and, after he had consolidated his control over the city, he sent some ships on the first of his exploratory missions south. By 1418 two of his squires had rediscovered Madeira, and his men were soon to re-explore the Azores and the Canary Islands, but in 1424 they reached Cape Bojador or the Bulging Cape, just south of the Canary Islands, and there they stuck.

In Henry's day, thanks to Ptolemy's *Geography* of the second century AD and other impeccable sources, sea captains knew three things: nobody could cross the equator because it was a wall of fire; nobody could sail round Africa to India because the Indian Ocean was an inland sea and Africa curled all the way round to Russia and Japan; and nobody could sail past Cape Bojador. People were not quite sure why Cape Bojador was impassable – some argued that its reefs and currents would dash ships to inexorable destruction, others suggested that it was the edge of the world over which ships would plunge – but all sea captains were agreed on the fact that to sail past Cape Bojador was impossible. Indeed, the waters around the cape are red, coloured by the desert sands as a warning from God, and the ocean currents strike the cape obliquely, generating ever more dangerous swells the closer it is approached. The shores of the cape itself in 1434, moreover, were desolate, unpopulated and sterile.

Between 1424 and 1433 Henry sent no fewer than fifteen expeditions to round the cape, and all fifteen returned with excuses of failure – they were the wrong kind of waves. Finally, in 1434 Henry sent Gil Eannes, who was not a professional sailor but a squire from his household, to brave the horrors, and as Eannes approached the dreaded cape he veered off westwards, sailing out to sea. When he turned south he found that Bojador was already behind him. Landing on the African shore he found no evidence of the gates of hell.

That was – as Francis Bacon noted – Henry's greatest contribution to humanity, because it showed that exploration could disprove received truths. As Bacon asked: 'Why should a few received authors stand up like Hercules columns, beyond which there is no sailing or discovering?'

Subsequently, Henry sent mission after mission to Africa, each creeping down the coastline a further 50 or so miles. Yet every time his men landed they found little but sand. In 1435 Eannes and Gonçalves Baldaya, some 50 miles south of Bojador, found footprints of people and of camels in the sand, but they found no actual people or camels. In 1436 Baldaya reached Cape Rio de Oro, 200 miles south of Bojador, but it was neither a river, nor inhabited, nor golden. In 1443 Tristão Vaz Teixeira reached Cape Blanco, 250 miles further south, where finally he discovered people on the shore; so in 1444 Eannes returned, bringing 200 of the people back to Portugal as slaves.

We know a lot about Henry because he was so admired by Gomes Eannes de Azurara (sometimes given as de Zurara; c.1410–74), the Portuguese chronicler who wrote the eulogistic *The Chronicle of the Discovery and Conquest of Guinea*. There Azurara described how tiresome African mothers could be, 'clasping their infants in their arms and throwing themselves on the ground to cover their bodies, disregarding personal injury to prevent their children being taken from them,' when the Portuguese were trying only to recover the costs of exploration.

Henry died in 1460, but Portuguese exploration did not die with him. In 1469 King Alfonso V, Henry's nephew, leased out to Fernão Gomes, a citizen of Lisbon, a five-year monopoly on the trade with the Guinea coast in return for a royal share of the profits and on condition that Gomes explored a further 300 miles of the coastline each year. It was a mutually profitable contract, Gomes discovering parts of Africa that were for centuries to be called, gratefully, the Pepper Coast, the Ivory Coast, the Gold Coast and the Slave Coast. Gomes even crossed the equator without being burned to death.

In 1474 Prince João, later to be King João II, took control of the exploration. But João soon faced a problem, namely that, south of the equator, navigation was difficult because the north star was no longer visible. But, helped by antisemitism in Spain, he lured to Portugal a series of Jewish cosmographers, including Joseph Vizinho, who in 1485 sailed to Africa to calculate latitude from the height of the sun at midday, and Abraham Zacuto, who, as the author of the authoritative *Almanach Perpetuum*, taught the Portuguese to navigate the southern seas. Soon João's captains were racing each other against different methods of navigation. In the words of Duarte Pereira, in his *Esmeraldo de situ orbis* (1505–08): 'Only when we reach the Cape shall we know who goes better, they by dead-reckoning or I by astrolabe.'

And in 1488 Bartolomeu Dias was swept by a storm around the Cape of Good Hope, and when he landed 200 miles to the north he found he was on the *eastern* side of Africa. He had reached the Indian Ocean! There was nothing now to stop the Portuguese breaking into the spice trade.

Except for the Spanish. For years Queen Isabella of Castille had been harrassed by a monomaniacal bore (and boor) called Christopher Columbus, a sailor from Genoa who, after having been shipwrecked off the coast of Portugal, had spent much of his life in Lisbon making charts. Columbus was obsessed with sailing west to Japan and from there to India. It was a good idea and, like many good ideas, not original. Even as Columbus lobbied the Spanish and Portuguese crowns, a German, Martin Behaim, was lobbying his own princes for a westward journey to India. Earlier, in 1474, Paolo dal Pozzo

Toscanelli of Florence had suggested the same westward mission across the Atlantic to King Alfonzo V of Portugal. And in 1487 two Portuguese, Fernão Dulmo and João Esteito, had actually sailed due west to see what they could find. They had found nothing.

But Columbus argued that Dulmo and Esteito had made the mistake of sailing west directly, in the face of the prevailing winds. He proposed first sailing south to the Canaries before turning west, because at that latitude the winds are easterly – that is, they blow from the east to the west. Yet Columbus knew that Dulmo and Esteito had not really made a mistake – he knew that they had deliberately sailed into the wind to ensure that, on turning, they could sail safely back home. Columbus, whose life's work as a mapmaker and navigator had gone nowhere, had nothing to which to return. He would risk all.

Intrigued by Columbus's certainties, King João II of Portugal had convened a committee to consider his plans, but the committee had dismissed them as unworkable: the world was too large for a sailor to reach Japan westwards from Europe. The committee was, of course, correct: Columbus would never have reached Japan. Nevertheless, Columbus lobbied other monarchs, not only the Spanish ones but also Charles VIII of France and Henry VII of England, but all spurned him.

Then in 1488 Dias returned to Lisbon from the Indian Ocean. Horrors! The Portuguese were about to capture the spice trade. After more study by more experts and after some tough negotiation with Columbus – Who was going to keep the profits from the trip? – Queen Isabella of Castile graciously changed her mind, entrusting him with the command of three caravelles, the *Santa Maria*, the *Pinta* and the *Nina*. On 3 August 1492 Columbus left Seville in southern Spain, a week later he left the Canaries, and on 12 October he heard the lookout cry *'Tierra! Tierra!'*

Undaunted by Columbus's successes to the west, the Portuguese continued to press on east, and in 1497 Vasco da Gama sailed out with four ships to cross the Arabian Sea and the Indian Ocean to India, a frightening trip across unfriendly and unknown waters. During the southern summer of 1497–8 da Gama crept cautiously up the eastern coast of Africa, encountering suspicious Arabs and antagonistic pilots in Mozambique and Mombasa before finally taking on board the famous Ahmad Ibn Majid, the author of *Kitab al-Fawa'id* (*Nautical Directory*) and the best Arab navigator of his day. It is believed that Ibn Majid kindly, and foolishly, piloted da Gama across to Calicut in southern India, the capital of the spice trade. The Portuguese had finally reached the subcontinent.

During the southern winter of 1498 da Gama tried to negotiate with the *samuri* (*zamorin*) or king of Calicut, pretending he was really only in search of Christians left over from Roman times, but the *samuri* was hostile, insulted by the cheap trinkets da Gama had brought as presents – no Indian king could have been impressed with gifts selected for African chiefs. After three futile months da Gama left for home, finally returning to Lisbon in September 1499. It had been a dangerous trip, with da Gama having lost two out of his four ships and 115 out of his 170 men, many to scurvy.

But in February 1502 da Gama set off again, this time in command of a significant fleet and this time in a militant mood. Gone was Vasca da Gama the polite explorer and searcher after lost Christians; in his place was Vasca da Gama the murderous conqueror, plunderer and pirate. Arriving off the coast of India on 1 October 1502, he encountered a large dhow, the *Meri*, carrying Muslim pilgrims home from Mecca. He sailed alongside, demanding the immediate surrender of their every possession. The pilgrims being slow to comply, in the words of one of da Gama's crew: 'We took a Mecca ship carrying 380 men as well as women and children, taking from it 12,000 ducats and goods worth another 10,000, then burning the ship and all the people who were on it.'

His Christian duty performed, da Gama sailed on to Calicut, ordering his old friend the *samuri* to surrender and to expel or kill all the Muslims. When the *samuri* tried to negotiate, da Gama rounded up some local Indians, took them back to his ship, hanged them, chopped their bodies into pieces and put the pieces into a boat, which he floated ashore with a large sign in Arabic suggesting that the *samuri* turn them into curry. The *samuri* yielded.

Thereafter the Portuguese conquest of the Indian spice trade was rapid. By 1509 Francisco de Almeida, the first Portuguese viceroy in India, had defeated the Muslim fleet, and by 1510 Afonso de Albuquerque, his successor as viceroy, had transferred the capital of Portuguese India to Goa. Albuquerque had already secured the Persian Gulf for Portugal by capturing Hormuz in 1507, and in 1511 he took Malacca in Malaysia, which gave him control of the major spice islands: every link of the spice chain, from cultivation in the Far East to sale in Lisbon, including reprovisioning in India and Africa, was now under Portuguese direction.

As early as 1503, morever, within a year of da Gama's second voyage, the Portuguese had already destroyed the Venetian monopoly over pepper, and its price in Lisbon that year was only one-fifth of that in Venice. And so profitable were the spices that by 1506, within four years of Vasco da Gama's second voyage to India, they had already increased the Portuguese crown's annual revenues by 25 per cent. By 1588 they accounted for more than half the government's income.

As Portugal monopolized the trade with the East that had once been Italy's and Arabia's, so the consequences were felt throughout Europe: within twenty years civic authorities the length of the continent were lamenting Portugal's rapacity as they had once lamented Venice's. A Nuremberg decree of 1523 bewailed that over 100 tons of ginger and over 2,000 tons of pepper were being imported annually into Germany from Lisbon and that 'the king of Portugal, with spices under his control, has set prices as he will, because no matter how dear they will be bought by Germans'.[1]

But not by the Italians. The citizens of Nuremberg never broke their addiction to their delicious spicy Nuremberg sausage, and there are still restaurants in Nuremberg that serve only sausage, but the Italians went the other way, foodwise. Having pioneered spicy food – it had been the Roman matron's boast that no guest could identify through the hot sauces the particular meat being served, and the Renaissance cuisine of Italy had been as hot and sauce-disguised – after 1503 the Italians, disgusted by their loss of the spice trade, embarked on the *nuova cucina*, inventing the light and simple style that, by concentrating on the taste of the ingredients themselves, was to transform cooking throughout Europe.

The rest of Europe, in short, was jealous and fearful of Portugal's and Spain's discoveries because those countries had grown so rich, bloated on the African trades in gold, slaves and ivory, the Indian trade in textiles, the Far Eastern trade in spices, the American trade in gold and silver, the Atlantic trade in cod and the other spoils of navigation. Other European countries, too, wanted to get into the game. But it needed to know how the game was played. The man who explained the rules was Francis Bacon.

FRANCIS BACON

Francis Bacon (1561–1626) was born in London in 1561. His parents were Sir Nicholas Bacon, the Lord Keeper of the Great Seal, and Lady Anne Cooke, who had studied Greek and Latin. Francis Bacon followed his father into the law, studying at Trinity College, Cambridge, but it was as a politician that he rose, becoming successively MP, Solicitor General, Attorney General, Lord Keeper (like his father) and Lord Chancellor.

Two of Bacon's great loves were power and money, and in 1621 he was found guilty of bribery, his estates were confiscated, and he was exiled in perpetuity from London. But it was those loves of power and money that inspired his scholarship because he was so fascinated by Spain's global dominance at the time of the Armada.

As we have seen, it was by his study of the Portuguese historians that Bacon concluded that Spain had acquired its power and wealth by copying Henry the Navigator. And how had Henry made his great discoveries? By scientific research. From the chroniclers Bacon learned that in 1419 Henry had retreated to Sagres, an isolated promontory in Algarve in southwest Portugal where, leading a celibate life of austere study and research, he had collected a group of geographers and astronomers and cartographers and shipbuilders to plan a systematic programme of scientific exploration. Under his direction, Henry's research group had improved the compass, developed the caravelle (a small, rakish ship with fore-and-aft sails and a large rudder that was especially manoeuvrable against the wind) and had constructed novel star maps and other navigational aids, including superior charts. Henry had created the science that had powered first the Portuguese and then the Spanish to global dominance.

In so doing, Bacon maintained that Henry had only copied the example of the Athenians, Ptolomies and other ancients who, 2,000 or more years earlier, in their academies, libraries, museums and other institutions of scholarship, had fostered the greatest science of their eras *and* the greatest power. Bacon wrote: 'For both in Egypt, Assyria, Persia, Graecia, and Rome, the same times that are most renowned for arms, are likewise most admired for learning . . . [they] . . . have a concurrence or near sequence in times.' (The *Advancement of Learning*, First Book, p. 11) Bacon thus concluded that Henry had confirmed that scientific research was the precondition for improvements in technology: 'If any man think philosophy and universality [science] to be idle studies, he doth not consider that all professions [technology] are from thence served and supplied.' (Second Book, p. 62.) It was therefore Bacon who first proposed the 'linear model' for economic growth:

$$\text{science} \rightarrow \text{technology} \rightarrow \text{wealth}$$

But who paid for science? Bacon said it was no coincidence that Henry was a prince and that the Ptolemies were pharaohs and that earlier researchers, such as Aristotle, had been funded by their rulers (Philip of Macedon in Aristotle's case). Bacon believed that science needed to be funded by the state because it was, in his words, a 'universality'. In his *Novum Organum* of 1620, Bacon said that: 'The benefits inventors confer extend to the whole human race' – that is, inventors benefit the *whole* human race, not any particular individual.

Today, we talk of 'private' goods and 'public' goods. Private goods are served satisfactorily by markets: a compass, for example, is a private good, because once a purchaser has bought it, the purchaser owns it, so why shouldn't the purchaser pay its full market price? But the *concept* that some metals are magnetic, and can therefore be used to navigate, cannot be individually owned: it is 'universal', and it benefits the whole human race; it is a public and not a private good. Consequently, no one will pay for its development because no one will pay for the development of a concept that cannot be monopolized but that will be used largely by others, including competitors, enemies and the unborn. So Bacon concluded (in an early claim of 'market failure') that for science 'there is no ready money', which was why governments had to pay for it. Bacon's full linear model therefore was:

$$\text{government money} \rightarrow \text{science} \rightarrow \text{technology} \rightarrow \text{wealth}$$

To propagate this idea Bacon in 1605 wrote *The Advancement of Learning* (which is still in print) to urge the British government to copy Henry the Navigator and to put money into university science.

OECD

In 2003, four centuries after Bacon wrote, a very different authorial entity, the Organization of Economic Cooperation and Development (OECD) published a book with a title that would have interested Bacon, *The Sources of Economic Growth in OECD Countries*. The OECD, like Bacon's thinking, was born of war, and it was created in 1947 as the allies' economic think tank. Today its member states are the richest market economies on the planet.

In its *Sources of Economic Growth* the OECD reported on a comprehensive regression analysis (248 pages of tables, figures and data) that surveyed all the known factors, ranging from the different national macroeconomic policies to the different national labour productivity decompositions, that might explain the different growth rates of the twenty-one leading economies of the world between 1971 and 1998. On page 84 the OECD found, unexceptionally: 'a significant effect of R&D [research and development] activity on the growth process' (that is, research and development drive economic growth). But then it found, explosively (and I quote four sentences in full, even though they are written boringly, because their contents are indeed explosive):

> Furthermore, regressions including separate variables for business-performed R&D and that performed by other institutions (mainly public research institutes) suggest that it is the former that drives the positive association between total R&D intensity and output growth

... The negative results for public R&D are surprising and deserve some qualification. Taken at face value they suggest publicly performed R&D crowds out resources that could be alternatively used by the private sector, including private R&D. There is some evidence of this effect in studies that have looked in detail at the role of different forms of R&D and the interaction between them.

The Sources of Economic Growth in OECD Countries, 2003

The OECD concluded, therefore, that it is business, not public, R&D that drives economic growth, and that the public funding of research and development displaces its business funding. Consequently, the public funding of research and development actually damages economic growth. Moreover, the OECD does not stand alone: at least two other researchers, Walter Park of the Department of Economics at the American University at Washington, D.C., and myself, have found – by similar surveys of OECD data – similarly damaging effects of the government funding of research and development.

This is an unexpected development. Ever since Francis Bacon said that governments needed to fund science, people have almost universally agreed with him, yet the economic think tank of the very governments that most fund science has now reported that their intervention is counterproductive. A cynic might wonder if the OECD was paving the way for governments to save money by withdrawing from science funding, but a cynic would be wrong because the science ministers themselves (certainly the ones I know) were as surprised as everybody else by the OECD's findings. Moreover, governments *like* funding science: the sums of money are relatively small, and politicians love to be associated with one of the few aspects of government expenditure that is measurably useful. No one who witnessed Bill Clinton and Tony Blair jointly celebrating the mapping of the human genome need doubt that, even though, ironically, the mapping was driven as much by two privately funded bodies, the Wellcome Trust and Celera, as by the governments in Washington and London.

WHAT IS SCIENCE?

Francis Bacon's *The Advancement of Learning* is now largely remembered not for its economics but for its creation of a new philosophy of science. In Bacon's day the Church and the universities taught that all knowledge had already been revealed and that further scientific truths were to be uncovered only by deduction from existing principles: an investigator seeking to uncover a particular fact – how far can this arrow fly? – was instructed to deduce it from Aristotle's laws. And scientists who deviated from deduction were persecuted. So the first great Bacon, Roger of Oxford, who had written in his 1267 *Opus maius* that 'arguments are not enough, it is necessary to check all things through experience', was imprisoned between 1277 and 1291 for 'suspected novelties'. In 1600 Giordano Bruno was burned at the stake for speculating about multiple universes (the earth couldn't be at the centre of all of them), and in 1633, for believing the sun to be at the centre of the universe, Galileo was shown the instruments of torture by Pope Urban VIII. Protestantism was no more liberal, and in 1553 Calvin burned Michael Servetus (who had discovered the circulation of the blood through the lungs) for questioning the doctrines of the Trinity and Baptism.

To handle the Church, medieval scientists resorted to one of the lasting euphemisms of history, pretending that 'search' was only 'research' and that 'discovery' was only 'recovery', and that, as in 'renaissance', they were only uncovering the old learning. But, as Bacon argued, they were doing no such thing. Not only had Henry's men rounded Cape Bojador in defiance of Ptolemy's *Geography* as early as 1434, but in 1530 Copernicus had written *De revolutionibis* to contradict Aristotle and put the sun, not the earth, at the centre of the universe. In 1543 Vesalius had written *De humani corporis fabricia* to contradict, by his own dissections, Galen's version of anatomy, and in 1600 William Gilbert from Colchester had written *De magnete* to transform – indeed invent – the sciences of magnetism and electricity (he coined the phrases 'electric attraction' and 'magnetic pole').

Inspired by these examples, Bacon honoured induction (the creation of new theories by personal observations) as the highest of achievements: 'Those who determine not to conjecture and guess but to find out and know . . . must consult only things themselves.' And he wrote *The Advancement of Learning* to provide a new description of science as a four-stage process of observation, induction, deduction, and experimentation.

SOME REVISIONISM

Bacon's description of progress and his invention of the discipline we call the philosophy of science were so influential that much of the European Enlightenment was forged in his image. Diderot and d'Alembert described him as the Father of their *Encyclopédie*, Voltaire called him the Father of Experimental Philosophy, and Immanuel Kant dedicated his *Critique of Pure Reason* to him. The Royal Society, meanwhile, was modelled on his prescriptions.

But 400 years is a long time, and much of Bacon's thinking has since required revision. We now, for example, recognize the 'problem of induction' (you can induce from observing every native British swan that all swans globally are white, and you would be wrong). Karl Popper showed us how to address that problem, namely by treating all scientific laws as provisional.

There is also a 'problem of observation'. Bacon believed that scientists collected facts dispassionately and then induced logically inevitable hypotheses. But scientists are actually like the rest of us. When witnesses are asked in court, say, to describe a traffic accident, they don't start by describing everything they did or did not see – 'I was wearing grey socks that day; there was no hot-air balloon flying overhead' – they describe what the cars were doing. As Einstein noted: 'It is theory which decides what we observe.' There is always subjectivity, judgement and creativity in science, and researchers select the observations they make.

And the OECD's findings have now suggested that Bacon got something else wrong, namely the funding of science. And since Bacon linked his philosophy of science with its economics, these errors of Bacon's force us also to refine our understanding of what science is.

Bacon believed that progress was the cumulative addition of discrete pieces of knowledge, each of which was a public good whose creation would be neglected by the market. But if the public funding of research and development damages economic growth, then science cannot consist of these discrete quanta of public knowledge. Science must be something else. In this book I show that science is not public but largely tacit and that consequently there is no such thing as science but only scientists.

To show these things I follow in Bacon's footsteps, chronicling the history and building a theory of science on it. But I have an advantage over Bacon: I have access to better history. Consider Francis Bacon's reverence for pure science. That was based on his readings of the Portuguese chroniclers, including Gomes Eannes de Azurara, João de Barros and Antonio Galvano. But we now know their books to have been biased. Indeed Barros ('the Livy of Portugal') proclaimed that he selected his facts to portray Henry and the Portuguese in a favourable light. And he was wise to do so: his friend Damião de Góis (1502–74) who was sufficiently distinguished to be also a friend of Erasmus's and to have his portrait painted by Dürer, wrote dispassionate history. But the Portuguese of the day did not appreciate dispassionate history, and Góis went to prison.

The problem for the chroniclers was that Henry was a brute – he abandoned, for example, his own brother to a lonely death in a Muslim prison – and a thug, who aspired to little but the killing of Muslims and the stealing of their trade. Sir Peter Russell's 2000 book *Prince Henry 'the Navigator': A Life* is only the latest revisionist destruction of the myth that Henry was a noble hero. Henry was not, for example, the spiritually inclined virgin of the chroniclers, but rather a commitment-phobe who produced at least one illegitimate child. He was not a scholar who maintained at Sagres an academic college of disinterested researchers; he was instead a professional soldier who employed technologists opportunistically, as cunning warriors do. And Henry's involvement with dispassionate science was negligible. Indeed, his so-called love of learning was, like his so-called virginity, invented by the chroniclers to camouflage his unpleasant nature.

And Henry had little love of exploration. He preferred killing people. He turned to 'navigation' in 1418 reluctantly, only because the Muslim armies' pressure on Ceuta was so strong that he could make no direct headway against them, so he tried to outflank them. But when, in 1437, that direct pressure eased and the opportunity arose of attacking Tangier, he promptly suspended his navigation and besieged the city for no fewer than five years. Only after the comprehensive failure of the siege did he, reluctantly, resume his navigation.

Not only did Bacon swallow Portugal's propaganda, he interpreted facts loosely. The 'concurrence or near sequence in times' for the arms and learning of 'Egypt, Assyria, Persia, Graecia, and Rome' that Bacon invoked to prove that science created power actually reflected the opposite: it was the arms (such as Philip of Macedon's) that came first and the learning (such as Aristotle's) second. The money and power came first, the knowledge second. As Philo of Byzantium (*c*.280 BC–*c*.220 BC wrote of Hellenic science: 'Alexandrian scientists were heavily subsidized by kings anxious for fame.' The money came first, the science followed.

As for Bacon's magic trio of gunpowder, the magnetic compass and printing, the first two had been developed in China by around the time of Christ, spreading to Europe via the Silk Road, while printing with movable type, though also developed by the Chinese, was independently invented in Europe by Johann Gutenberg, a goldsmith, around 1440. In none of these cases are the inventors believed to have been anything other than artisans or traders. They were not scholarly researchers.

* * *

Henry the Navigator was a major figure, who not only launched Europe to world domination but who also put western Europe on top of eastern Europe. For millennia Europe was eastern-directed: in 330, for example, the Emperor Constantine moved the capital of the Roman Empire to Constantinople, reflecting the eastern balance of an empire whose most populated province was Egypt, and it was Constantine who promoted Christianity as Europe's dominant religion – yet that faith originated in Asia, as did Jewry and Islam, and of the five patriarchs of the early Christian Church only one, the pope in Rome, was based in the western part of the empire, the others being in Constantinople, Alexandria, Jerusalem and Antioch.

But four millennia of eastern Mediterranean hegemony were to be ended by Henry's transfer of the spice profits to Lisbon, which presaged the long-term transfer, from the countries bordering the

Mediterranean to the countries bordering the Atlantic, of European vitality in every field of human activity. The new worlds were to speak English, French, Dutch, Spanish and Portuguese, not Greek, Italian or Arabic.

The Italians resisted decline, of course. By 1530, having renegotiated with the Arabs, the Venetians had recovered some of the spice trade and were importing over 1,000 tons a year. Though that amount was still dwarfed by Portugal's 3,000 or more tons, the spices were fresher and therefore better priced. But the spices, like the gold from Senegal, were for the Portuguese only a catalyst for further trades. Slaves were as profitable as spices and gold, and, once the Portuguese had mastered long-distance navigation in search of slaves, spices and gold, they expanded into the even more valuable trades of Newfoundland cod, Brazilian sugar and many others. The Portuguese had been driven to master long-distance navigation only by necessity – that mother of invention – but thereafter their invention powered them to ever-greater wealth, and no last-minute trade renegotiations could compensate the Italians for their technological obsolescence.

As the commercial advantage moved to the Atlantic seaboard, so the great bankers would no longer bear Romance names, such as Bardi, Medici, Acciaiccioli or Peruzzi, but northern ones, such as Schroder, Hambro, Kleinwort, Peabody, Morgan or 'Jew Rothschild, and his fellow Christian, Baring' (Byron, *Don Juan*). And, since science follows wealth, soon the great scientists would no longer bear Romance names such as Eustachius, Vesalius, Fabrizio, Fallopius and Galileo, but northern ones, like Copernicus, Kepler, Leibniz, Newton and Harvey.

All this was precipitated by Henry the Navigator. Earlier I wrote of his driving his men to round Cape Bojador as a great *cultural* advance, because it challenged the intellectual rigidity of his day, but it was also a great geopolitical and commercial advance. Henry, in short, may have been a bad man but he was surely a great one ('Great men are almost always bad men,' Lord Acton).

We are today familiar with Europe's recent history of world domination, but the Europeans' rise was not inevitable, and to contemporaries it would have seemed an absurd proposition. As late as 1776 in his *Wealth of Nations* Adam Smith could write that: 'China is a much richer country than any part of Europe . . . The retinue of a grandee in China or Indostan accordingly is, by all accounts, much more numerous and splendid than that of the richest subjects in Europe.' The economic historian Paul Bairoch has confirmed that, in Adam Smith's day, China still indeed accounted for 33 per cent of the world's manufacturing output, and India 25 per cent, when Europe could account for only 23 per cent. GDP per capita, moreover, was higher in the great empires of the east than it was in Europe.[2]

But the pendulum was swinging, and by 1900 China could account for only 4 per cent and India 2 per cent of the world's manufacturing output, whereas Europe and its white diasporas accounted for almost all the remaining 96 per cent and for most of the world's income and wealth too. And it was Henry the Navigator who gave the pendulum the nudge.

Francis Bacon was surely right, therefore, to identify Henry as a pivotal figure and equally right to identify his use of technology as seminal, but because Bacon did not have access to the history we have, he invented a linear model of science as a public good that was a myth. Unquestionably Henry was a great researcher, but he was not the researcher of Francis Bacon's model. In this book, therefore, I show that science cannot be the public good of Bacon's vision and that actually to understand science we have to write the history afresh.

I start with a review of the centuries leading up to 1688, the date of the Glorious Revolution in England, which was the event that launched the Industrial Revolution. When we've reached 1688, I shall review what the history has told us about science, and I shall show that the history of the Industrial and Post-Industrial Revolutions confirms those lessons. I shall then review what the economics tells us about science. So let us start at the beginning . . .

Book 1
Early History

1
Stone Age Good ...

'The Stone Age did not die because we ran out of stones.' Old joke

In 1836 a Danish archaeologist, Christian Thomsen, named much of human history for technology: the Stone Age, the Bronze Age, the Iron Age. Technology is not restricted to humans – birds build nests, beavers make dams and spiders spin webs – but those animals' behaviour is programmed; creative technology is only a few millions of years old, and it is produced largely by ourselves and by our three nearest living relatives, the gorillas and the common and bonobo chimpanzees. Once, people believed that only we humans used tools intelligently, but in her 1971 book *In the Shadow of Man* Jane Goodall reported that the chimpanzees use sticks and stones creatively as, moreover, do gorillas. Jane Goodall's report was unexpected, and it was not welcomed by those who believed in the uniqueness of human intelligence, but it has since been comprehensively confirmed. Indeed, we now know that Jane Goodall's report was not even original: in 1906 the Post Office in Liberia issued a 5 cent stamp illustrating a chimpanzee using a stick to dig termites out of a mound.[1]

It is not necessarily surprising that we share tool-making with the African great apes because the evidence from fossils and from DNA shows that we four ape species separated from each other only some seven million years ago, the proverbial flicker of the eye in terms of evolution. We humans share between 95 and 98 per cent of our DNA with the chimps.[2] We did not, of course, actually evolve from those three apes, because they have been doing their own evolving over the

31

last seven million years: the four species have been separately evolving from our common ancestor. We do not know if that common ancestor used tools, because the earliest tools the archaeologists can identify are stones only some two and a half million years old (so the Stone Age is formally two and half million years old), so we do not know if the four species each independently learned to use tools over the last 7 million years or if they inherited the ability from our common ancestor. Only when the DNA and neurological networks that underlie tool-using have been characterized for all four species will we know if they evolved independently or commonly.

PROGRESS

Even though the Stone Age lasted at least two and half million years, progress during it was slow, and until 100,000–200,000 years ago our protohominid ancestors were only hunter-gatherers or foragers. They were nomads, and the animals they killed (as judged by the bones they left) were only little ones. Humans did not kill dangerous mammals such as buffaloes or wild pigs; they were so primitive that, though they often lived by the sea or by rivers, they may not even have fished. About 100,000 to 200,000 years ago, however, men and women started to advance technologically, a period that archaeologists call the transition into the Upper Old Stone Age. People learned to make clothes out of animal skins, they learned to make needles and thread, and they finally learned to fish, developing harpoons and hooks from bones. They even mastered fire.[3]

We cannot be sure why people finally started to accelerate technologically around 100,000 to 200,000 years ago, but the anthropological evidence, as collated in books such as Steven Pinker's *The Language Instinct* (1994), shows that articulated speech seems to have developed around then. Recent molecular research on FOXP2, the first characterized gene for speech, confirms that it too is 100,000–200,000 years old.[4] Language is an information technology, so we have here the first evidence in support of Bacon's ideas: new technology emerged as information spread. Knowledge is power.

By 70,000 years ago we humans had made our last biological advance when we invented art, which means that we had, by then, developed aesthetic thought (the earliest known art forms seem to be the stone ochre carvings and seashell necklaces from the Blombos Cave in South Africa, dated to some 70,000 years ago[5]). And because we know from DNA studies on the Y chromosome that our last common male ancestor or 'Adam' lived some 60,000 years ago ('Eve' may have been older, but that does not change the story), we can complete the biological story of our species, namely that some 70,000 years the last wave of humans – who were also the best equipped intellectually – emerged from Africa (or possibly the Middle East) to colonize the globe.[6]

Modern *Homo sapiens* or Cro-Magnon (named after the cave in France where bones were first found) reached southern Europe around 50,000 years ago, where we found the Neanderthals in possession. So we drove the Neanderthals into extinction – either by direct battle or by successfully competing for the same resources – and, in a forerunning of the propaganda that such conflicts spawn, we have long rubbished the Neanderthals as brutes. Yet though their stone tools and weapons were more primitive than ours, we are belatedly recognizing that Neanderthal brains were slightly larger than ours, that the Neanderthals buried their dead, and that they cared for their sick. But they had no art, so they were presumably less capable of aesthetic thought (and they would have been hairy, as well as unattractively short and stocky), so perhaps it was all for the best.

It was about 10,000–15,000 years ago that people really took off technologically, in what the archaeologist V. Gordon Childe called the New Stone Age Revolution. This was *the* important step: it generated civilization. (Childe, one of the most influential of archaeologists, was a Marxist, so he used the term 'revolution'. That term is now increasingly passé in intellectual circles, but I still find it useful, just as I prefer AD and BC as more honest dates than BCE and CE. Indeed, I find BCE and CE offensive; why should we Europeans assume our Gregorian milestones to be common to all?)

Different peoples embarked on their New Stone Age revolutions at different times. The first one, and the one that was ultimately to breed European, Middle Eastern and Indian civilizations, started in the area of the Middle East known as the Fertile Crescent, roughly coinciding with modern-day Iraq, Iran, Jordan, Israel and parts of Turkey. There, about 11,000 years ago, women and men invented farming – it is often said, though no one really knows, that it was the women who first developed farming. People settled down, planted crops, grazed animals and built houses; and because, for the first time, these agricultural humans had homes, we get the word 'domesticate' from *domus*, the Latin for home.

WHAT WAS THE NEW STONE AGE?

The New Stone Age is generally presented as a story of technological progress. The Middle East, after all, was a good site for an agricultural revolution because it was rich in species for women and men to domesticate. There were wild plants, including wheats, barleys and peas, to collect, and they are good to eat. They are, though, hard to harvest. The bits we eat are the seeds, which are full of energy, but those plants did not package them for our consumption, they packaged them for dispersal: when a wild wheat or barley plant is ripe

34

its stalk shatters, dispersing the grain, and when a pea plant is ripe its pod explodes, dispersing its seeds. But occasionally a mutation occurs, and individual wheat and barley stalks do not shatter and individual pea pods do not explode.

In the wild such mutations die out for obvious reasons, but for humans they are a joy because they make the collection of seeds and peas so much easier. And the archaeologists have found that, around 8,000 BC, those mutations were being collected by our ancestors. Today, across the globe from America to Australia, the world's farmers still largely cultivate the progeny of those original Fertile Crescent mutants, as well as the olives, lentils, chickpeas, figs and flaxes that also emerged from the Middle East.

The domestication of the animals will have followed related lines. Tribal peoples across the world today domesticate an extraordinary variety of young animals, ranging from baby kangaroos in Australia to baby otters in the Amazon rainforest. It would not be unreasonable to suppose that the peoples of the Fertile Crescent will have, similarly, domesticated lambs, kids, piglets and other local baby fauna. As those creatures matured, so people will have learned to husband them until they were ready for the pot.

The first species to be tamed was the dog which, some time before 10,000 BC, started to live alongside us. People were still largely nomadic, still hunting and gathering, and it made sense for the dogs, who could roam alongside, to join up; we got help with the hunting, they got a share of the kill. Later, around 8000 BC as the tribes started to settle, less mobile species such as sheep, goats and pigs were tamed. By 6000 BC the Egyptians had domesticated the cat (which they worshipped) and the donkey. The Egyptians were good with animals, taming cheetahs, hyenas, giraffes, gazelles, hartebeests, eagles and cranes. (Few animals, incidentally, domesticate easily. To domesticate, a species must have evolved to live collectively, so it's used to company; it must be hierarchical, so it does as it is told; and it must be good-natured. Not many animals meet all these criteria – the zebra, for example, is apparently similar to the horse, but it has a horrible personality and it bites.)

The New Stone Age was not just an agricultural revolution, it was also an industrial revolution, and the peoples of the Fertile Crescent developed a host of impressive skills, including spinning, weaving, dyeing, leather working, distilling, the grinding of corn and the baking of clay for pottery. Tool manufacture, stone working and the beginnings of architecture also emerged.

Meanwhile, from about 7500 BC, the peoples of China (rice, millet and silkworms) and India (sesames, aubergines and cows) were embarking on their own technological revolutions, and by 3000 BC the native Americans were harvesting maize, beans, potatoes, turkeys, llamas and guinea pigs. The Europeans were not especially innovative, mostly importing their domesticates from the Fertile Crescent, but they did develop poppies and oats. As Dr Johnson noted, the Scots eat oats.

That is the conventional story of the New Stone Age – a heartening tale of technological progress.

WHAT WERE NEW STONE AGE PEOPLE LIKE?

There is, also, a conventional story to account for the creativity of the Stone Ages, though the story has changed over the last century. Obviously we will never know precisely what New Stone Age people were like because they left no written records, but the archaeology is expressive, and some tribal peoples still survive and, presumably, resemble the tribals of a few thousand years ago.

A century ago the West supposed the surviving tribal peoples to be stupid and treated them casually. When in 1897 Robert Peary, the explorer who first reached the North Pole, transported some Inuit or Eskimo people to the US, he displayed them as living exhibits in the American Museum of Natural History.[7] As recently as 1904 a pygmy man from Africa was being exhibited in the Brooklyn Zoo, sharing his cage with an orang-utan[8] – and only lately has the museum in Banyoles, Spain, removed its stuffed African from show.

But books such as Claude Lévi-Strauss's *The Savage Mind* (1962) have corrected these misapprehensions, and today we understand the tribals to be no more 'backward' than other members of our species – they have simply inherited different cultures. The surviving tribals generally inhabit isolated territories or land that does not readily support agriculture, so they may not have become farmers, but, as they reveal by their comprehensive knowledge of their environments, the tribals are no stupider than the farmers. Consider their botanical knowledge: when during the 1980s Bruce Chatwin visited Australia to research his book *Songlines* he met a female ethnobotanist who was interviewing 'Old Alex', an aboriginal, and Old Alex's knowledge of the local plants could not be exhausted. He knew the name of every plant the ethnobotanist had collected, he introduced her to dozens more, and he knew where each could be found, when each flowered, which ones were edible, which ones were positively tasty and which were poisonous.

Just as the surviving hunter-gatherer peoples are walking encyclopedias of natural history, so too were the early foragers of the Fertile Crescent. An archaeological excavation of the charred plant remains at Tell Abu Hureyra in the Euphrates Valley in Syria (10,000–9000 BC) showed that its peoples brought home at least 157 different species of plant.[9] Yet these were not randomly collected: no toxic or poorly nourishing wild species were represented. Other archaeological excavations have found further evidence for the wide

SEX, SCIENCE AND PROFITS

yet discriminate collection of wild plants by the early hunter-gatherers globally.

We can see, therefore, that hunter-gatherers are alert to their surroundings, that they are informed, curious, and – most importantly – that by testing new plants they experiment. Tribal cultures will, of course, have differed in their openness to trial and error, and even today the Ibos of Nigeria, the Navajo of North America and the Jews of Europe remain more entrepreneurial than some of their neighbours, but we can see that many of the hunter-gatherer peoples will have been as questioning and as intelligent as ourselves. They could have powered a New Stone Age.

WHY A NEW STONE AGE?

But why did the New Stone Age start some 10,000–15,000 years ago? Why not 100,000 years ago – or never? In his 1951 book *Man Makes Himself* V. Gordon Childe suggested that the New Stone Age Revolution had been precipitated by climate change. He suggested that the Ice Age of 10,000 or so years ago had so disrupted hunting that humans had been forced into farming. But that particular Ice Age was merely the twenty-third in the series, so why had the earlier ones not forced us into farming?

Other explanations have been offered for the introduction of agriculture, including simple opportunism – 'Oh look, the seeds we dropped accidentally last year have produced a crop' – but the most credible – and grimmest – explanation for the New Stone Age, one that is now widely accepted, was provided during the late 1960s by two remarkable archaeologists, Lewis Binford and Kent Flannery.[10] They noted that the revolution had been preceded and accompanied by the extinction of no fewer than 200 or so animal species. They also noted that the New Stone Age Revolution of the Middle East appears to have been grim for the people living through it: the human skeletons that have been excavated from New Stone Age times are smaller, less nourished, more diseased and younger than those from the preceding Old Stone Age, suggesting that the food supply during the New Stone Age was poorer than during the old one.

Instead of representing a Great Leap Forward, therefore, Binford and Flannery concluded that the New Stone Age Revolution appears to have developed in response to starvation. In one study Mary Stiner, Ofer Bar-Yosef and their anthropological colleagues collected animal bones from Stone Age sites in the Nahal Meged in modern-day Israel and from the western coast of Italy.[11] By restricting their studies to bones that had been charred or cut, the anthropologists knew they were looking at the remains of animals that had been butchered and cooked by humans. And the anthropologists found that 100,000 or so years ago our ancestors ate tortoises, hedgehogs, shellfish (especially limpets), ostrich eggs and other slow-moving creatures.

But by New Stone Age times, 10,000–15,000 years ago, those easily caught animals were largely absent from our diets – they had been hunted to near-extinction – and the rare specimens that were still being eaten were smaller than before, reflecting their scarcity and our hunger. Instead, our consumption of rabbits, hares, marmots, squirrels and other faster moving prey increased as we were forced to work ever harder for our food. And when, eventually, even that small prey become scarce, and when even the bigger prey such as wild gazelles became scarce, the pressure on resources forced humanity to farm – or starve.

We may, therefore, be able to explain why the New Stone Age started some 10,000–15,000 years ago. Life for the Old Stone Age hunter-gatherers had been good but, during the Upper Stone Age the development of speech so improved our hunting skills that by

10,000–15,000 BC we had hunted most of our prey into extinction. We may never know exactly what happened 10,000–15,000 years ago, but the most likely explanation is that, simply, we were hungry.

MASS EXTINCTIONS

It is now widely accepted that we humans have precipitated mass extinctions. Some of the most recent were perpetrated by the Indonesians. It was the Indonesians who grew many of the spices that Europe so loved and who, incredibly, traded them by carrying them 4,000 miles across the Indian Ocean to East Africa in nothing more substantial than double outrigger canoes, before sending them up the coast to Egypt via the Arabs of the Red Sea. This ancient spice route has left its traces. The island of Madagascar, for example, was uninhabited when the Indonesians discovered it, which is why the language of Madagascar to this day, even among the later African and Indian immigrants, is a dialect of the Ma'anyan language of Borneo.

But the Indonesians also provoked an ecological tragedy. Because Madagascar had been isolated from the mainland for millions of years, a remarkable local fauna of giant elephant birds, giant lemurs, pygmy hippopotamuses and other bizarre animals had evolved there, but the Indonesians hunted them to extinction, and now we know those animals only from their skeletons. More recently the Indonesians were as murderous to the now-extinct moas of New Zealand and the flightless geese of Hawaii when they got there in AD 500 and 1000, respectively.

Earlier, some 40,000 years ago, humans reached Australia, whereupon they killed off the giant kangaroos, diprotodonts or giant marsupial rhinoceroses, marsupial leopard-like animals, 400 pound ostrich-like birds, one ton lizards, giant pythons and giant crocodiles and other unique animals of that island continent (though I cannot entirely condemn the aborigines of Australia for their treatment of the giant pythons and the terrestrial giant crocodiles). Only a minority of researchers still blame climate change, not human activity, for the mass extinctions.[12]

North America is believed by most scientists to have been colonized only some 13,000 years ago, perhaps because men and women had to wait for the end of the last Ice Age before the Bering crossing could be made – whereupon they promptly killed off the herds of mammoths, mastodons (elephants) and horses and camels and giant ground sloths grazing the grasses under the interested gaze of lions and cheetahs and smilodons (sabre-toothed tigers). Romantics have argued that the mass extinctions of America were not caused by early people but by the last Ice Age, but why did the preceding twenty-two Ice Ages, unlike the one that probably coincided with human immigration, not kill off the big animals of North America? It was Paul Martin who, in his paper 'The Discovery of America' published in 1973 in *Science*, demonstrated how the early American humans had both the opportunity and the motive to kill them off.

As they did in the Fertile Crescent. People arrived, and first they hunted the slow prey into near extinction and then the small fast prey, and then the big, fast prey, such as the wild gazelles, and suddenly the Crescent wasn't so fertile anymore. Faced with starvation, men and women turned for help through the technological innovations of the New Stone Age agricultural revolution. But *why* did we hunt the animals into extinction?

PUBLIC GOODS AND THE TRAGEDY OF THE
COMMONS

Imagine that you are a tribesperson of the Upper Old Stone Age some 100,000 years ago. Life is good, your men are killing lots of animals, your women are producing lots of babies, and your people are healthy and growing in numbers. But over the horizon are different tribes – different humanoid species even – with nasty foreign languages and nasty foreign habits. And their nastiest habit is that of popping over the horizon on occasion and killing your wild animals.

'Your' animals? Er, no. Nobody owns them, actually, and sometimes they themselves migrate over the horizon, where you face another of the alien tribes' nasty habits – their resistance to your killing of 'their' animals. What to do? Obviously you kill as many foreign tribespeople as possible, yet that's dangerous because they kill back. Or you try to make peace with them, yet that's also dangerous because you can't be sure they'll keep their side of the bargain so it's best to get your retaliation in first. Regardless, it's best to produce the maximum number of children to outbreed the enemy, and it's best to kill the maximum number of animals so that your tribe eats well and their tribe starves.

You have, in fact, set yourself up for a classic 'tragedy of the commons' as described by the naturalist Garrett Hardin – the situation where, if no one owns a common good because it is public, everybody competes to exploit, and so exhaust it, first.[13] You have, in fact, set yourself up for the mass starvation that precipitated the New Stone Age Revolution. Here, therefore, is the modern story of the New Stone Age – a story of 'universalities' or public goods.

THE SEX LIVES OF THE NEW STONE AGE
REVOLUTION

Some people reject starvation-via-overhunting-and-overbreeding as an account of the New Stone Age because they observe the tribal peoples who still survive. Tucked away in corners of rainforests, isolated on tropical mountain plains or remote on marginal land, many of these tribal peoples have remained close to the original Stone Age in their foraging lifestyles, and these surviving tribal peoples do not necessarily overbreed, nor do they exhaust the local animals. It is no romantic myth, it is simply true, that many tribal peoples live in equilibrium with the local animals and plants.

So, do tribal peoples commune spiritually with Nature and with the animals, Dr Dolittle-style? If so, they do it with uncommon cruelty, because the extant tribal peoples seem not to be tender-hearted or troubled by animal rights. Tribal practices can be routinely and casually cruel to animals. But Lewis Binford, the archaeologist who proposed the New Stone Age Revolution to have been driven by starvation, has also shown that the extant tribal peoples live in equilibrium with the animals – not because the peoples are kind but because they tend to live in isolated or self-contained environments. Consequently extant tribals, in practice, exercise collective ownership over their animals, and to husband those animals the tribals will limit their population growth.[14]

Where tribals do not limit their population growth, it is because their environment is so hostile that their survival is threatened not by overpopulation but by the sheer hostility of that environment. When the naturalist Redmond O'Hanlon was researching his 1983 book *Into the Heart of Borneo* he found that life in the highlands of Borneo was so dangerous for the tribals and their death rates from accidents, disease and predation were so high that they enjoyed a vigorous sex life. Repopulation was as much a duty as a pleasure – one that the youngsters, O'Hanlon reported, undertook conscientiously.

But, in one of the great discoveries of modern anthropology, we now know that extant tribal cultures living in less hostile lands limit their population growth. Infanticide, for example, has been common. In their 1988 book *Homicide*, their global review of murder, Martin Daly and Margo Wilson, the husband-and-wife team of McMaster evolutionary psychologists, showed that more than half of all human societies have selectively killed babies, often for economic reasons. Here is a report from the anthropologists Kim Hill and Magdalena Hurtado of an interview with an Ache man. The Ache are a small group of foragers who live in the rainforest of eastern Paraguay:

> The man told his story with tears welling up in his eyes and explained that it was the Ache custom to kill children after their parents died . . . but the killer explained that 'the old powerful men told us we had to kill all the orphans'.
>
> Hill and Hurtado, *Ache Life History*, Aldine de Gruyter,
> New York, 1996

Some societies have killed children systematically. The Spartans of ancient Greece exposed their newborn boys on hillsides, and their death rates were so high that, famously, much property in Sparta was owned by women. The Romans, on the other hand, were enjoined under their oldest laws, the Laws of Romulus, to raise all their male children but to retain their first-born daughters only; the other girls they abandoned at the Lactaria Column. ('If you give birth to a boy, look after it – but if a girl, let it die', Hilarion wrote to his wife in 1 BC.) Among the tribals certain Inuit (Eskimo) tribes culled their baby girls, which is why they institutionalized polyandry (many husbands to one wife), and, until stopped by the British, some indigenous tribes in Australia ate their surplus children.

Contraception, too, is common among the tribals. Bronislaw Malinowski, the famous anthropologist from the London School of Economics, wrote in his 1929 book *The Sexual Lives of Savages in North-Western Melanesia* that 'savages' did not understand that sex led to babies. Really? In her 1980 book *Sex in History* Reay Tannahill showed how universal contraception has actually been, ranging from Onan spilling his seed on the ground to the Zulus engaging in *uku-hlobonga* (a form of *coitus interruptus*) to the Carrier Indians of British Columbia, who were one of many peoples who developed menstruation taboos to limit procreation, exiling their pubertal women to episodes of seclusion in the wilderness. Contraception is so old that the Kahun Gynaecological Papyrus of around 1825 BC recommended the insertion of crocodile dung within the vagina to inhibit conception. The Romans, meanwhile, developed abortion, the Arabs castrated their African slaves, and sodomy was practised extensively in classical times.

Birth control has, in short, been exercised for so long and so widely that Adam Smith described it as a Law of Population in his *Wealth of Nations*: 'the demand for men necessarily regulates the production of men' – that is, Smith suggested that people have only as many children as they need. He noted, for example, that: 'A half-starved Highland woman frequently bears more than twenty children, while a pampered fine lady is often incapable of bearing any, and is generally exhausted by two or three.' Smith noted that the poor had more children because their children were more likely to die: 'It is not uncommon in the Highlands of Scotland for a mother who has borne twenty children not to have two alive.' (Both quotations take from *Wealth of Nations*, Book 1, Chapter 8).

Modern scholarship supports Smith's observations that prosperous societies limit their populations. In 1700, for example, the average Englishman did not marry until the late age of twenty-eight, and his average bride was twenty-seven. Moreover, parish records show that age on marriage depended on wealth; in hard times people married late, in good times early.[15] In those days, of course, extramarital intercourse was rare, with fewer than 2 per cent of mothers being unmarried, so age on marriage reflected fecundity.

If Smith confirmed that people, no matter how primitive, can match their populations to demand, let us ask why, if the original Stone Age could have regulated its fertility, it didn't?

OWNERSHIP

The New Stone Age was a terrible time because nobody owned the animals on which they depended, so people hunted them to extinction, and every advance that humanity made, either in hunting technology or in tribal organization, aggravated the problem, because the more that people bred or hunted in their personal interests, the worse the situation developed collectively. We see here, therefore, how the search for individual self-interest can be collectively self-defeating. This echoes Francis Bacon's belief that self-interest cannot sustain 'universalities'.

One solution to the New Stone Age problem was provided by agriculture, not because it provided food as such – food can always be exhausted – but because it provided a food source that could be owned. A resource that is owned will be husbanded, not extinguished. Farmers who have sowed will not destroy their crops or exhaust their seed corn in a desperate race against others – they will protect and grow their crops. People may have discovered only by chance that seeds could be planted and harvested, but once they had sown the seeds they would not have doubted to whom the crops belonged. They had, in fact, invented the concept of ownership. The human invention of ownership represents one of the major inventions of history, because it made human history possible. It was the private ownership of crops that reversed the 100,000-year decline in Upper Old Stone Age living standards and that provided the economic surplus off which civilization was to feed, literally.

Regardless, therefore, of how agriculture originated, its perpetuation was self-reinforcing because people who own their land and crops and animals will not surrender them lightly.

* * *

Francis Bacon was therefore right to identify ownership as key to development. Much of the Old Stone Age descended into starvation, and the people who survived were either those who invented agriculture (with the property rights that accompanied it) or who, as isolated tribals, effectively owned their local flora and fauna. And with ownership came the incentive to limit population. With it, too, came the incentive to develop technology, because the person who owns their means of production has an incentive to improve them.

But wait: didn't Bacon also say that technology flowed out of pure science, which was a public good that needed government money? So how did the new ownerships of the New Stone Age promote technological growth? An early answer to that question was provided by Adam Smith.

2

Adam Smith Biologist

'Man, an animal that makes bargains.' Adam Smith

Adam Smith (1723–90) lived 150 years after Francis Bacon, and, like Bacon, he understood that technology was the basis of economic growth. But unlike Bacon he believed that governments got in the way: only markets, Smith said, fostered useful technology.

As a young man Smith had studied philosophy at Glasgow University, then a good institution, but he moved for further study to Balliol College, Oxford University, which was not a good institution. At Glasgow Smith had met eminent scholars, such as Francis Hutcheson the philosopher, Joseph Black the chemist, and eventually James Watt of steam engine fame. But at Oxford there was . . . nobody. As Smith was to write: 'In the University of Oxford the greater part of the public professors have, for these many years, given up altogether even the pretence of teaching.' Smith's contemporary at Magdelen College, Oxford, Edward Gibbon the historian, was to write: 'My tutors were monks who supinely enjoyed the gifts of the founder [endowments]. My own [tutor] well remembered he had a salary to receive, and only forgot he had a duty to perform. [My] fourteen months at Oxford were the most unprofitable and idle of my whole life.'

These experiences inspired Smith to seek an explanation for the superiority of the Scottish over the English universities, which he found in the market. The Scottish universities had to earn their money by fees, whereas Oxford (and Cambridge, then no better) fed off their endowments: 'Improvements were more easily introduced into some of the poorer universities which were obliged to pay more attention to the current opinions of the world.' Academics at Glasgow and Edinburgh, unlike those at Oxford or Cambridge, were paid by the students, so the lecturers had to please those students – and different academics competed to teach similar courses to the same students. But Oxford and Cambridge limited entry, admitting as lecturers only unmarried men who were ordained in the Church of England. The average age of an Oxford or Cambridge academic then was around twenty-seven, and most were only waiting to be appointed to a vacant parish, where they were allowed to marry. So, for example, the Rev. Dr John Michell FRS, who first postulated black holes as early as 1784, resigned his Cambridge post for a parish in Yorkshire when he wanted to get married. Such moves were not likely to foster scholarship, though it was by watching the young fellows wait for their livings that Smith conceived of his Law of Population.

Smith extrapolated from his academic experiences to conclude that markets, not governments or cartels or central planners, delivered wealth, and he largely lived by his beliefs. When Glasgow University, unusually, offered him a regular stipend in recognition of his inter-national reputation, he refused, saying: 'A teacher's diligence is likely to be proportioned to the motive which he has for exerting it.' Smith was cynical about people who sought government grants for providing public goods: 'I have never known much good done by those who have affected to trade for the public good . . . people of the same trade seldom meet together, even for merriment and diversions, but the conversation ends in a conspiracy against the public.' Indeed, Smith was cynical about 'that insidious and crafty animal vulgarly called a statesman or politician'.

So in 1776, the year of revolution, Smith published *The Wealth of Nations* to explain that governments should do practically nothing but keep the peace, protect private property and enforce contracts. Governments did not generate wealth, only entrepreneurs did.

SMITH'S ECONOMIC MODEL

Smith had faith in markets not only because they provide economic incentives but also because they provide a mechanism for technological growth: specialization. Smith was famous for his story of the pin factory: men who work alone might make only some twenty pins a day, but if they specialize – with one worker smelting the metal, another putting on the blunt bit at one end, and another sharpening the other end and so on – then productivity shoots up 240-fold to 4,800 pins a person a day.

Smith believed that specialization was key to an even more important phenomenon, namely research and development (R&D). Smith believed that innovation came from specialists working on the factory floor, not from academics in university labs. Smith told of how, when touring a factory, he once met a boy who was employed to regulate the valve on a 'fire' or steam engine during each phase of its cycle, and who then built a link between the handle of the valve and the shaft of the piston to open and close the valve automatically. By extrapolating from that anecdote into a systematic survey of industry, Smith concluded that Bacon was wrong to suppose that industrial advance emerged out of science: actually, most industrial advances emerged from workmen on the job: 'A great part of the machines in manufactures were initially the inventions of common workmen who naturally turned their thoughts to finding easier and readier methods of performing their work.' (*Wealth of Nations*, Book 5, Chapter 1)

And Smith, who, unlike Bacon, was both a university teacher, and had personal experience of an industrial revolution, suggested that

Bacon was also wrong to suppose that academic science led industrial technology; actually, the flow went the other way, with academic science feeding off the advances made in industry: 'The improvements which, in modern times, have been made in several different branches of philosophy [science] have not, the greater part of them, been made in universities [that is, they've been made in industry].' Here is Smith's model, which is very different from Bacon's:

academic science ← new technology → wealth

↑

industrial money + old technology

Because Smith's belief that science flows out of, not into, technology, may seem counterintuitive, let me offer a modern example, radioastronomy. The science of radioastronomy – that apparently purest of sciences – emerged during the 1930s when Karl Jansky (1905–50), an engineer working on long-distance radiotelephony for Bell Telephone Laboratories, a commercial outfit, discovered a source of electromagnetic 'noise' as coming from the stars. Out of that industrial finding a whole academic discipline was born.

But radioastronomy provides but one of innumerable examples of science emerging out of technology because simple history reveals that science *has* to be the daughter of technology: technology came so much earlier. The first proto-scientists were the priest-astronomers of Sumeria who appeared around 3000 BC, at the very end of the New Stone Age.

HOW DID THE NEW STONE AGE REVOLT?

If the New Stone Age had no scientists, how did it produce its new technology? By markets, of course. The archaeologist Colin Renfrew and his colleagues have shown that, over 6,000 years ago mollusc shells from the Black Sea were being traded, presumably as currency, over hundreds of miles across the Balkans and central Europe of the New Stone Age.[1] New Stone Age trade routes for many products extending from obsidian to amber to flints to axe heads, and ranging over hundreds of miles from specialized areas of production or mines or salt pans or fields or pastures, have been found in all the major continents. The New Stone Age thus specialized, and over 6,000 years ago some tribes were already professional miners, toolmakers or salt pan evaporators, whereas others were farmers or sailors.

Within recent times, before their indigenous economies were disrupted by Western intrusion, tribal peoples sustained their ancient patterns of trade. The Trobriand islanders of Papua New Guinea inhabit an environment whose interior is infertile but whose periphery is rich in fish and in productive soil. So they have traded: the inhabitants of the interior manufacture pots and hardware in exchange for food from the periphery. (Some tribal peoples sustain unexpected specializations. In his 1951 book *Transformation Scene: The Changing Culture of a New Guinea Village* Ian Hogbin reported that some groups in New Guinea survive by producing only . . . poetry! This they exchange for food and materials. Today's assumptions that art needs government subsidy might not survive a survey of the Stone Age any better than do the assumptions about science or technology.)

The evidence of the New Stone Age, moreover, suggests that technology is not a public good. Consider the spread of agriculture into Europe. About 10,000 years ago agriculture spread into Europe from the Fertile Crescent, and scholars have asked if it was *farming* that spread as the hunter-gatherers learned to copy their neighbours – that is, farming is a public good – or if it was *farmers* who spread, killing and displacing the locals – that is, farming is a private good. Thanks to the DNA population studies of L.L. and F. Cavalli-Sforza, which have shown that about 20 per cent of the average European's DNA is Middle Eastern in origin, we now know that it was the farmers who migrated, taking their new technologies with them.[2] Thus, during the New Stone Age, knowledge did not behave as a public good that could be accessed easily by others; it behaved like something embodied or tacit or tribal or private. There was then no such thing as farming, there were only farmers.

It would be absurd, of course, to draw strong policy prescriptions for today solely from the New Stone Age, but nonetheless we can conclude that the New Stone Age – the most important industrial revolution in human history – was an age of trade, not of government-funded universities (which, of course did not then exist) and that it progressed by Smith's model, not Bacon's: the great centres of the New Stone Age Revolution were the plains of the Fertile Crescent, India and China, because they could sustain markets large enough to reward technological innovation.

THE BIOLOGICAL BASIS OF TRADE

According to Smith, therefore, specialization fosters R&D. And what, in turn, fosters specialization? Markets. Markets foster specialization because markets allow specialists to survive by bartering their speciality products for food, clothes and the products of other specialists. Smith believed, therefore, that once markets flourish, specialization will also flourish, leading to R&D and thus economic growth.

In his 'other' great book, *The Theory of Moral Sentiments* (1759), Smith considered what markets really were. He concluded that they were a form of human cooperation. Smith said that 'man has almost constant occasion for the help of his brethren', and he proposed that trade evolved to foster that help. To put it in today's language: 'I'll give you this if you give me that, and we'll both be better off.' Consequently, Smith said, trade was an inherited instinct: 'The propensity to truck, barter and exchange one thing for another . . . is common to all men.'

Later studies have supported Smith's suggestion that humans are instinctive traders. In his 1839 book *Journal of Researches into the Geology and Natural History of the Various Countries Visited by HMS Beagle* (known usually as *Voyage of the Beagle*) Charles Darwin described how 'primitive' were the natives of Tierra del Fuego, yet 'they had a fair idea of barter. I gave one man a large nail (a most valuable present) without making any signs for a return; but he immediately picked out two fish, and handed them up on the point of his spear.'

Few, if any, human tribes have adjoined without bartering with each other. They may have fought, but they've bartered too. It was the anthropologist Napoleon Chagnon who, studying the Yanomamö peoples of the Venezuelan rainforest, whose separate villages are forever fighting each other, noted how villages sometimes pretend to 'forget' how to make certain objects such as pots to provide an excuse for trade and so for peace.[3] Trade is universal, and when my infant children went through their *Pokémon* stage (now thankfully passé) I discovered how correct Smith was in supposing it to be instinctive.

We can build on Smith's insight by considering the evolutionary roots of trade. Consider cooperative group behaviour among animals. That is not obviously based on trade; rather, it seems to be based on hierarchy, the so-called pecking order. The details vary (in some species leadership is shared among males, while in others, such as the African hunting dog, a female may lead), but generally a group of animals is led by an older male, and he decides where to go, what to do and, as compensation for the burdens of office, with whom to sleep. His subordinates obey him. This hierarchy can be called the pecking order. First described in 1913 by Thorlief Schjelderup-Ebbe, a Danish zoologist who was studying hens (hence its name, hens peck each other, female fowls having their own hierarchies), it is found in all social vertebrates, which makes it some hundreds of millions of years old.

In his classic 1949 book *King Solomon's Ring* Konrad Lorenz described the establishment of pecking orders in adolescent jackdaws. He and other scientists, studying a wide range of species, have reported that the establishment of a pecking order is fairly consistent among vertebrates; it's a peer group thing. Each generation of youths fights for status. Soon certain individuals start routinely to win, and others to lose, and a hierarchy is established.

Once everybody knows who's boss, the status fights become redundant – the losers learn to anticipate defeat and cease to challenge their subordinate position. Peace reigns, and hierarchy thereafter is signalled by gestures, with grooming or bowing or the presenting of bottoms substituting for fights. Gestures are important: if individuals were forever fighting over rank, forever drawing blood, they would soon drive each other into extinction; it is the substitution of gestures for real fights that has allowed vertebrate sociability to evolve.

But hierarchy is actually a contract, like a human barter. When a dominant animal snarls at his subordinate he threatens to tear him from limb to limb, unless the subordinate makes obeisance such as presenting his bottom to be mounted in faux coition. Yet if the subordinate does make that obeisance, the dominant animal is then pacified. A contract, where obeisance is bartered for acceptance, has been made between the two individuals.

A good contract between humans satisfies both parties: so it is between animals. Consider the vervet monkey, which Michael

McGuire of the University of California has studied. McGuire has shown that when subordinate vervet males make their gestures of submission, the dominant ones experience a surge of good feeling, which McGuire could chronicle by studying their postures and by tracking their brain levels of a chemical called serotonin.[4] And even subordinates, on making their gestures of submission, experience a relief from anxiety.

If the vertebrate pecking order is a contract, with one animal trading obeisance for another's acceptance, then the evolution of human barter, which is a contract with objects instead of gestures being traded, is easy to understand. Indeed, one 'missing link' between the gesture exchanges of the animal hierarchy and the object exchanges of human trade can be seen among nursing mothers, since the mother/child contract represents the trade of a gesture (crying) for an object (milk).

Nearer to home, the sex life of chimpanzees seems to close another of the missing links between the gesture exchanges of the hierarchy and the object exchanges of trade. Chimpanzees are apes, not monkeys, and – horribly enough, as Jane Goodall discovered – they like to eat monkeys.[5] But only the male, not the female chimpanzee, hunts for meat. Thus the chimps have set up one precondition for barter, namely specialization, it being the male chimps that obtain the monkey meat. Now, what do female chimps possess that male chimps might fancy in exchange?

Yup, you've got it. Meat is a private good for chimpanzees, while a female's acceptance of sex is a private good for males. Since barter comprises the exchange of private goods, it is no surprise that male chimps will provide females with meat in exchange for sex. This represents the barter of an object (meat) for a posture (female acquiescence in coition). And even human males provide their females with food as a prelude to sex: Sheila Sullivan's 2000 literary survey

Falling in Love showed that 98 per cent of seduction scenes in fiction were preceded by a meal, *Tom Jones*-style.

Chimps have even evolved the full object/object of a barter contract: if a band of male chimps kills a monkey, it will allow other males from the same group some meat under 'tolerated theft', which, however, requires a reciprocal offer of meat from the recipients next time they've successfully killed a monkey.[6] Chimps have therefore evolved a full promissory object/object barter contract on the human model. The evolution of human trade may not have been too different.

BARTER IS FEEDBACK

If barter is so central to the human condition, why of all the metaphysical possibilities does it take its particular shape? To us a contract seems 'natural', but is there any reason for contracts to adopt any special form? Would trade between aliens on Alpha Centauri look different from our own?

Let us consider, from the perspective of the dominant animal, the structure of a dominance/subordinate contract. The dominant animal (i) sees a subordinate, (ii) threatens that subordinate, (iii) extracts obeisance from that subordinate, (iv) feels better in consequence and (v) metaphorically pats that subordinate on the head.

Now consider, from the perspective of a buyer, the purchase of shoes. The purchaser (i) sees an attractive pair of shoes, (ii) offers money for it, (iii) receives the shoes, (iv) feels gratification and (v) puts them on.

Now consider the actions of a toilet cistern when it is being flushed: (i) it is flushed, (ii) the cistern ball falls, releasing the stopcock, (iii) water rushes in, (iv) the ball rises, closing the stopcock and (v) that arrests the flow of water.

In the engineering speciality known as control theory, the most important principle is that of the 'feedback loop'. Feedback loops provide automatic regulation, as the toilet cistern illustrates. Flushing activates the loop by causing the cistern ball to fall. But after the water that has been admitted by the stopcock raises the ball and closes the stopcock, so the loop also closes. The ball-valve flushing toilet is not, therefore, just a piece of plumbing, it is also an autonomous regulation system that feeds back information to modulate itself. It was invented, incidentally, in 1778 by a young Englishman called Joseph Bramah.

So consider, from the perspective of control theory, the structure of a dominant/subordinate or shoe-buying contract. The dominant animal or shoe purchaser (i) sees a subordinate or shoe and initiates the loop by (ii) offering a threat or money, (iii) he or she then extracts obeisance or a shoe, (iv) he or she is gratified, and (v) he or she pats someone on the head or slips on the shoes. It's a feedback loop.

Feedback is the basis of the nervous system. When you extend your arm you are (i) activating nerves to (ii) move your muscles, but (iii) as those muscles move they activate other nerves, which (iv) feed back the information on the whereabouts of the muscles to the brain, which (v) modulates the first set of nerves. Nervous systems cannot operate without feedback, and their feedback systems are over a billion years old. DNA, the chemical basis of life, may be over 3 billion years old and, as François Jacob and Jacques Monod, the French Nobel laureates showed, that too is regulated by feedback.[7]

Our contracts, therefore, are based on simple control theory, and because the feedback loop is the simplest biological expression of control theory, we can assume that life on Alpha Centauri will also be modelled on it, and that trade between intergalactic aliens will adopt the same metaphysical structure as our own.

* * *

We thus see that, over two centuries ago, Adam Smith proposed that trade was the instinct that underpinned cooperation and that trade, in its turn, fed technological growth because it fostered specialization, it being specialists who perform research to improve their technologies. And Adam Smith denied Bacon's claims that technology flowed out of science or that research needed government subsidies – rather, he proposed that industrial competition underpinned innovation.

The New Stone Age would appear to to have vindicated Smith's, not Bacon's, vision, but rather than dally any further, let's ask what the next chronological era, the Bronze Age, tells us.

3

. . . Bronze Age Bad

'When a man tells you he got rich through hard work, ask him
"Whose?"' Don Marquis (1878–1937)

One afternoon in Egypt some 4,500 years ago, around 2620 BC, the
Pharaoh Snefru complained of being bored. Demanding distraction,
Snefru summoned his chief of ceremonies, who suggested a cruise
down the Nile, a cruise with a twist. He suggested the royal barge be
crewed not by the usual male slaves but by the most beautiful girls in
the palace: 'You will see them row and you will see the river banks, the
fields and the bushes.'

According to the Westcar Papyrus, our record of that afternoon's
entertainment, the pharaoh approved of the plan. He called for twenty
oars of ebony and gold, and he called for twenty pretty girls, young
ones who had not yet lost their figures through childbirth. He wanted
twenty girls with perfect breasts, but he did not want them to display
those breasts, he was more subtle, and he found he was happy
contemplating them as they rowed.

We can still see those girls, or at any rate their sisters in servitude,
depicted on the walls of the thousands of tombs that have been
excavated in Egypt. Thanks to the cultural imperialism of nineteenth
century archaeologists we can also see them in the great museums of
New York, London, Paris and Berlin, and they reveal that the royal
palaces of the Old Kingdom resembled nothing so much as Hugh
Hefner's pad in Chicago, being peopled by armies of near-naked slave
girls who wore little but slim garters round their slim hips.

The higher ranking women, the queens and daughters and sisters and mothers of the nobility, were allowed to cover themselves more modestly, though even their linen gauzes could be surprisingly revealing, but for the slave girls there was no protection from the gaze of the pharaoh. Indeed, as the Westcar Papyrus explains, Snefru commanded the rowers to be specially covered for the cruise down the Nile. He was sated, and he needed the titillation of mystery. Not that there was too much mystery, for Snefru could have had any of those girls any time, any way, any place. He owned a vast harem of them, and, if the funeral painters of ancient Egypt can be trusted, the pharaohs had excellent taste. Their slave girls were lithe, honeyed and the possessors of the sweetest little breasts. Hugh Hefner, with his more obvious preferences, would have commanded matters on a grosser scale.

Floating down the Nile on his royal barge, Snefru would have passed ordinary Egyptians on the riverbanks doing ordinary things. Some of them, training to be architects, might have been sitting exams in maths. We know of some of the questions they were set. Here is one: 'A building ramp is to be constructed 730 cubits long, 55 cubits wide, containing 120 compartments filled with reeds and beams, 60 cubits high at its summit, 30 cubits in the middle, with a platform of 15 cubits and a pavement of 5 cubits. How many bricks will be needed?'

The specimen answer required quadratic equations. Over 4,000 years ago, therefore, the Egyptians had already mastered intermediate algebra – no trivial achievement, and they knew it. The Egyptians were proud of their intellectual and technological advances, and they had deified Imhotep, the physician who had designed the first pyramid, the Step Pyramid. The Pharaoh Snefru, when he was not enjoying slave girls, was to build three pyramids of his own, including the Rhomboid; and his son Cheops (Khutu), inheriting his father's monumental interests, was to build the Great Pyramid and the Sphinx.

But there were to be no further architectural developments in Egypt, no more innovatory Imhoteps.

Some 1,300 years later, around 1300 BC, another pharaoh, Rameses II, would float down the Nile on his own royal barge. One thousand three hundred years is a long time, yet, as Rameses sailed along, he would have seen little different from Snefru. From the design of his barge to his view of the early pyramids, Rameses's world was effectively the same as Snefru's. Other than the building of more pyramids and temples, the essentials of Egypt's urban, agricultural and ecclesiastical landscapes went unchanged between 2,600 and 1,300 BC, and if Rameses had read the examination questions in maths being set for architects in his day, he might have found this question: 'A building ramp is to be constructed 730 cubits long, 55 cubits wide, containing 120 compartments filled with reeds and beams, 60 cubits high at its summit, 30 cubits in the middle, with a platform of 15 cubits and a pavement of 5 cubits. How many bricks will be needed?'

Snefru and Rameses ruled over the richest country of their day, and their wealth was fabulous. The power of Egypt was fabulous too, and for thousands of years Egypt survived as an independent, often commanding, state, a state whose power was based on its mastery of technology. The Egyptians could not just calculate, they could irrigate the fields, they could breed crops and animals, they could write, they could build boats. So rich was Egypt that it could sustain a population of 7 million; so sophisticated was Egypt that it could retain that then enormous population as one political unit. The Egyptians were a Bronze Age empire when bronze was at the cutting edge of technology. But they did not develop. Why not? Why did the New Stone Age – which was such an exciting age – produce as its successor such a stagnant age as the Bronze Age?

THE BRONZE AGE

The Bronze Age might appear at first sight to have been exciting, because it was the first urban age or Age of Civilization (from *civitas*, the Latin for city), being the age of the first towns and cities. It was around 5500 BC, in the Fertile Crescent, that the early farming communities coalesced – or were forced – into larger political units or chiefdoms of some thousands or tens of thousands of people: capital cities, temples, castles, streets and other monuments appeared.

And, initially, technology flourished. Indeed, the early urban civilizations made such advances in irrigation and dam construction that in *Oriental Despotism: A Comparative Study of Total Power* (1957) Karl Wittfogel described many of them, including the Egyptian, Sumerian, Assyrian, Chinese, Indian and Mexican, as the 'hydraulic' civilizations for their remarkable control of water. They also developed boats and sails. And they worked metals. Copper was first identified in eastern Anatolia, in modern-day Turkey, around 6500 BC, though it did not become really useful until around 3000 BC, when people learned to alloy it with tin to make the bronze for which the age was named.

Meanwhile, the early Bronze Age invented writing (*c.*3500 BC in Sumeria, *c.*3000 BC in Egypt; *c.*2500 BC in the Indus Valley), and in a classic demonstration of Adam Smith's contention that technological advances emerge out of the market, it was the merchants who invented writing, to inventorize their goods. As Joan Oates confirmed in her 1986 history *Babylon*: 'Writing was invented in Mesopotamia as a method of book-keeping. The earliest known texts are lists of livestock and agricultural equipment. These come from the city of Uruk.' At the same time and for the same reason, the merchants also invented mathematics. And the Babylonians invented the wheel. Neither the Indus Valley nor China, which developed writing at around the same time as Egypt, was far behind technologically.

SEX, SCIENCE AND PROFITS

Coinciding with these advances, the first recognizable states of hundreds of thousands or of millions of people were being created. The Sumerians or Babylonians in modern-day Iraq led the way around 3600 BC, but some 500 years later the Pharaoh Menes was founding the first dynasty in Egypt, and about 500 years after that the people of the Indus Valley were building their own great cities of Mohenjo-Daro and Harappa (which survive today in present-day Pakistan as well-preserved ruins of baths, granaries, temples, court-yards, streets laid out in grids, and systems of water supply and of sewers that seem positively modern). But thereafter everything went dead.

IMPERIAL SEX

To understand what went wrong during the Bronze Age, we might look at its sex life. Although the different peoples of the different parts of the world embarked independently on their different civilizations – the historian Arnold Toynbee identified six independently developed civilizations: the Sumerian, Egyptian, Minoan, Chinese, Mayan and Andean – all six were to breed similar empires characterized by patho-logical snobbery, racism, cruelty and economic stagnation. Behind the oh-so-civilized city walls, inhumanity flourished.

Although the empires differed from each other in detail, they were sexually similar. We know of Pharoah Snefru's collection of sex slaves, but he was just a template, because all the emperors engaged in sex on an industrial scale. Consider the Chinese. The *Records of the Rites of the Chou Dynasty*, which were written around AD 1000, reveal that:

The assistant concubines, eighty-one in number, share the imperial couch nine nights in groups of nine. The concubines, twenty-seven in number, are allotted three nights in groups of nine. The nine spouses are allotted a night, the three consorts are allotted a night, and the empress is allotted a night. On the fifteenth day of the month the sequence is complete, after which it repeats in reverse order.

Quoted by Daniel Boorstin *The Discoverers*, Random House 1983.

The Bronze Age emperors enjoyed vast harems. In her 1986 book *Despotism and Differential Reproduction: A Darwinian View of History* Laura Betzig of the University of Michigan chronicled some of them: the Inca sun king Atahualpa, maintained 1,500 women, Hummurabi of Babylonia several thousand, the Pharoah Akhenaten 317 plus thousands more consorts, the Aztec Montezuma 4,000, the Indian Udayama 16,000 and so on and on. The aristocrats of these empires also maintained harems, though of carefully diminishing size as social status fell, and a junior official of the Inca Empire had to make do, poor chap, with only three wives. Still, that was three more than most Inca men saw, though not as many as did the king of Dahomey. That promiscuous monarch from West Africa claimed *all* of the women of the kingdom as his, and only on a whim would he release some to his favourite courtiers and warriors.

Nor was S&M neglected by the imperial era. The Mayan temples in Central America show reliefs of kings passing lengths of string through their pierced penises while their queens tortured themselves by passing lengths of string through their pierced lips (oral). The Inca, Aztec and Mayan ceremonies of murderous ball games and of human sacrifices, moreover, speak of institutionalized cruelty.

And the *mores* of empire could be openly erotic. The Indian temples are monuments to erotica, and their statues of big-bosomed naked girls fornicating with naked men – the best carvings are in Khajuraho – are still arousing, even in these jaded times.

IMPERIAL COMMAND ECONOMIES

The empires clearly exploited women, and they exploited men, too. They were absolute hierarchies. The spirit of the Egyptian Empire was encapsulated in a surviving papyrus in which an official dictates to his subordinates: 'No one has the right to do what he likes because everything is managed in the best possible way.'[1] That was, admittedly, a late or Hellenistic papyrus, but it captured the entirety of the imperial experience, and the similar spirit of the Sumerian empires was summarized in a surviving and much older cuneiform tablet: 'Without compulsion no settlement could be founded, the workers would have no supervisor, the rivers would not flood.' The antique empires were therefore command, not market, economies. And they were profitable for the emperors. Here is a passage from Genesis 47:24: 'Ye shall give the fifth part unto Pharaoh, and four parts shall be your own, for seed of the field, and for your food.' One-fifth of the farmers' production went, therefore, to the pharoah, which was *why* he ruled his empire.

It was the anthropologist Robert Carneiro who in his paper 'A Theory of the Origin of the State' explained that it required agriculture to create empire.[2] Men have always fought, but they started to create empires only 10,000, not 100,000 or 1,000 years ago, because only with the onset of agriculture did empires make economic sense. The fighting of the old Stone Ages did not lead to empire because the people of the old Stone Ages were nomads, and if defeated they melted away. But once dependent on agriculture, defeated warriors cannot melt away. If farmers are to eat they have to stay on their land. So, if defeated, farmers will have to collaborate with their victors – to the satisfaction of those victors.

Entrepreneurs during the Bronze Age, therefore, soon appreciated that the exploitation of the peasantry provided them with the easiest source of wealth, and exploit they did. Around 300 BC the Indian poet Valmiki, the supposed author of the two great Sanskrit epics *Mahabharata* and *Ramayana*, lamented that agriculture had destroyed the Golden Age of foraging and hunter-gathering:

> In the Golden Age, agriculture was an abomination. In the Golden Age people lived on fruits and roots that were obtained without any labour. But after the fall into the Silver Age, impiety appeared in the form of agriculture. For the existence of sin in the form of cultivation, the lifespan of people was shortened.
>
> *Ramayana*

Within the empires nobody enjoyed economic rights; there was no private property. Indeed, within the empires nobody enjoyed any human or legal rights. In the absence of an independent judiciary not even the emperors and priests possessed rights as we understand them, only privileges. Ordinary folk, therefore, were disposable. Consider the problem of personal service for the emperors after death: would there be palace staff in heaven?

This question worried the emperors. Habituated to service from the moment they were born, they expected the same from the moment they died, so, from the beginnings of the Bronze Age in Sumerian times many of the kings, queens, priests and aristocrats of the imperial era took their slaves and courtiers with them to the other side. In some cultures the slaves and courtiers were buried alive, in others they were slaughtered first, in yet others they committed ritual suicide, but in all cases their fates were secondary to those of the masters and mistresses they were to attend in perpetuity. And in some cases their numbers were considerable: one of the death pits of Ur, discovered by the archaeologist Leonard Woolley in 1923 and dated to around 2500 BC, contained the bodies of sixty-eight female court musicians.[3] Another contained a platoon of soldiers. Here is an account of age-old Balinese funerary practices recorded in 1820 by John Crawfurd in his *A History of the Indian Archipelago*. He is describing how, on the death of a prince, his female slaves were prepared for their post-mortem duties:

> Some of the most courageous demanded the poignard [dagger] themselves, which they received with their right hand, passing it to the left, after respectfully kissing the weapon. They wounded their right arms, sucked the blood which flowed from the wound, and stained their lips with it, making a bloody mark on the forehead with the point of the finger. Then returning the dagger to the executioners, they received the first stab between the false ribs, and a second under the shoulder blade, the weapon being thrust up to the hilt towards the heart. As soon as the horrors of death were visible upon the countenance, without a complaint escaping from them, they were permitted to fall to the ground.

IMPERIAL SNOBBERY

When the Portuguese reached India during the fifteenth century they found it a land of rigid caste distinctions. The caste system was originally racist. The word caste derives from the Portugese *casta*, meaning lineage, which in turn came from *casto*, meaning pure or chaste. The Vedic or ancient Hindu word is *varna*, which means colour, reflecting the social hierarchies imposed by light-skinned invaders from the north. The age-old Indian obsession with colour can still be seen in today's marriage advertisements, in which people ask for partners of a 'wheaten' complexion. These marriage advertisements are so coded that it is hard for Westerners to understand them, but 'wheaten' is clear enough.

Caste rules are divisive, separating people into priests (brahmins), warriors (kshatriyas), peasant farmers (vaishyas), serfs (shudras) and untouchables, or Dalits or scheduled castes (pariahs). Members of the higher castes are traditionally polluted if even the shadow of a lower caste person falls on them or their food, and they have to undergo ritual purification and reject the food. Yet this social system has, until recently, so dominated Indian culture that even Mahatma Gandhi, the great liberation hero, started his autobiography *My Experiments with Truth* with the words 'The Gandhis belong to the Bania caste'.

But the caste system was like an orthopaedic or plaster cast: it immobilized society and people. It certainly crushed trade. Consider navigation: why, in Henry the Navigator's day, did the Indians not trade their own spices? Why, indeed, did many Indians not travel? They did once. Bali, in Indonesia, is 95 per cent Hindu today because Indian traders settled there from around the first century AD. And the Buddhist text *Jatakas* describes how Indian sailors used 'compass-birds' to navigate; like Noah in the Bible, the sailors would release a crow, which would then fly to the nearest point of land.

But during the fifth century AD, under the Gupta dynasty, the caste laws began to be tightened, and eventually many Hindus stopped travelling. The discouragement of travel among some Hindu groups was so profound as to last into contemporary times. In Chapter 12, 'Outcaste', of his autobiography, Mahatma Gandhi described how as a young man during the 1880s he had wanted to study in Britain, but no member of the Modh Bania caste had ever travelled abroad, so a general meeting was called. Addressing it, the *sheth*, or head of the community, proclaimed: 'Our religion forbids voyages abroad. It is not possible to live in England without compromising our religion, one is obliged to eat and drink with Europeans!'

The antique empires discouraged travel both because travel exposes people to new ideas and because travel, by facilitating trade, empowers merchants and other non-officials. In both China and Japan, for example, it was for centuries a capital offence even to speak to foreigners. Thus did the emperors minimize commerce, reducing it largely to the importation of their personal luxuries. In Edward Gibbon's words: 'The objects of oriental traffic were splendid and trifling.' (*Decline and Fall of the Roman Empire*, 1776, volume one, chapter two). And even that residual trade the emperors and their priests monopolized, organizing it from their palaces or temple for their sole benefit.

LOSING GROUND

The imperial age, therefore, represented an age of loss – and not just of trade or of quadratic equations. The imperial loss was also obvious artistically: the New Stone Age cave paintings of Lascaux provide more honest and more vivid representations of reality than do the repetitive and formulaic wall paintings of Egypt or the bas-reliefs of Assyria. The imperial loss, moreover, was obvious intellectually. In his 1988 *Hierarchy, History and Human Nature: The Social Origins of Historical Consciousness* the University of California anthropologist Donald E. Brown reported that, of twenty-five different civilizations, none organized as hereditary castes fostered objective history, they floated only on seas of myths and legend.

Today it is easy to marvel over the pyramids and other imperial wonders and to overlook that the people of the Bronze Age had lost the human and economic freedoms of the Stone Ages. They certainly lost their human, property and trading rights. Barter may be an instinct, but instincts can be crushed by tyannies. Trade and private ownership, therefore, can flourish only if a society is sufficiently equitable.

While we cannot pronounce with certainty about equity during the

Stone Ages, many of the surviving tribal societies seem to have been reasonably equitable. In *Voyage of the Beagle* (1839) Charles Darwin described how the natives of Tierra del Fuego lived in a state of 'perfect equality . . . even a piece of cloth is torn into shreds and distributed; and no one individual becomes richer than another'. Christopher Boehm, the anthropologist who directs the Jane Goodall Research Center of the University of Southern California, noted that among contemporary tribal peoples:

> The Kalahari San 'cut down braggarts'. Among the Hazda, 'when a would-be chief tried to persuade other Hazda to work for him, people openly made it clear that his efforts amused them' . . . Australian aborigines 'traditionally eliminated aggressive men who tried to dominate them'. In New Guinea, 'the execution of a prominent individual who has overstepped his prerogatives is secretly arranged'.
> Boehm's quotes from E. Sober and D.S. Wilson, *Unto Others: The Evolution and Psychology of Unselfish Behaviour*, Harvard University Press, 1998

The tall poppy clearly has deep roots. Boehm has concluded that the anthropological evidence suggests that:

> as of 40,000 years ago, with the advent of anatomically modern humans who continued to live in small groups and had not yet domesticated plants and animals, it is very likely that all human societies practised egalitarian behaviour.

But Boehm's research also suggests that humans will abandon their egalitarianism and adopt oppressive hierarchies, 'insofar as they work for the common good'.[4] And such a common good did emerge – that of killing and eating one's neighbours. The fearsome tragedy of the commons of the last 40,000 or so years, as the different tribes of the plains of Eurasia and the Americas fought over ever-diminishing food, rewarded those people who forsook their easy-going egalitarian ways for fierce, disciplined warfare, culminating in the Bronze Age tyrannies where the development of agriculture rewarded the oppression of the hapless many by the armed few.

The loss of human, trading and property rights, therefore, led to a loss of innovation because, in the absence of trade, individuals lost the incentive to research into new technologies and because, in the absence of property rights – without the security of actually owning anything – people will not improve things that will only be stolen from them. The imperial age, therefore, was sterile; but that sterility suited the emperors just fine. If you were Snefru, drifting on your barge of slave girls, would you want to change very much?

Now we see why the extra food and surplus wealth of the New Stone Age did not translate into further progress. The food and wealth of the first agricultural and industrial revolutions were not retained by the workers or traders, who might have invested it in some useful technologies. Instead, it was creamed off by kings, priests and soldiers to support their palaces, temples, pyramids, weapons, barges and slave girls and to crush dissent or heterodoxy. Even Francis Bacon, an admirer of Hellenistic Egypt, lamented that: 'among the Egyptians, who rewarded inventors with divine honours and sacred rites, there were more images of brutes than of men; inasmuch as brutes by their natural instincts have produced many discoveries, whereas men . . . have given birth to few or none' (*Novum Organum*, 1620).

The great empires were not ignorant. During the seventh century BC, Ashurbanipal, the last great Assyrian emperor in Mesopotamia, boasted that: 'I solve complex mathematical reciprocals and products with no apparent solutions, and I read abstruse tablets whose Sumerian is obscure and whose Akkadian is hard to construe.'

Ashurbanipal, who collected a considerable library of cuneiform tablets, explained that he had been universally acclaimed from his earliest youth as the cleverest member of his generation. But none of that Assyrian learning translated into progress.

* * *

In her book *Despotism and Differential Reproduction: A Darwinian View of History* Laura Betzig of the University of Michigan argued that the Bronze Age empires could best be understood biologically: the dominant males formed coalitions to oppress all other males and to monopolize the females. The dominant males thus propagated their genes preferentially. Betzig's is an intriguing story, and it may even be true, but it's not necessary to invoke it. It's enough simply to say that entrepreneurs are rational, and that if they live in a society where wealth can be stolen more easily than it can be created, then they will cooperate to steal it.

The empires of the Bronze Age did understand ownership – the emperors owned everything and everyone – but they did not recognize individual property rights or free markets, and consequently they failed to innovate. To create wealth requires a different sort of society – one that could create an Iron Age.

4
The Iron Age Invents Science

'O ferrum! Heu ferrum!' – 'Look at the iron! Alas, the iron!'
Desiderius of Lombardy bewailing the glinting armies besieging
him in Pavia, 773

In a famous speech delivered soon after the outbreak of the First
World War in 1914 David Lloyd George, who was to lead Britain to
victory four years later, said:

> This is the story of two little nations. The world owes much to little
> nations. The greatest art in the world was the work of little nations; the
> most enduring literature of the world came from little nations; the
> greatest literature of England came when she was a nation the size of
> Belgium fighting a great empire. The heroic deeds that thrill humanity
> through generations were the deeds of little nations fighting for
> freedom. Yes, and the salvation of mankind came through a little
> nation.
>
> Quoted in B. Macarthur, *Penguin Book of Historic Speeches*, 1996

Lloyd George could have been describing humanity's escape from the
Bronze Age. Had the whole world then succumbed to the imperial
armies, humanity would have stagnated into totalitarian sterility, but
on the periphery of the great empires were sometimes to be found little
nations and, if those little nations were also traders, then art, literature,
heroic deeds and, most importantly, technical innovation emerged.

We know some of the little nations from the Bible. Thus it was the

Canaanite merchants who, during the second millennium BC, needing a proper script to record transactions, developed the first alphabet. Bronze Age writing had been hieroglyphic (each word having its own symbol) and therefore burdensome, but the Canaanite invention of the alphabet was to transform Mediterranean literature and, transmitted and modified by the Phoenicians, Greeks and Romans, the letters of this book were originally Canaanite.

Writing harnessed economic power and, people being attracted to wealth, the words grammar and glamour share the same roots. Now that everyone can write, grammar's relative economic power has been devalued, and in these philistinic days it's the semi-literate super-models who are glamorous. In Canaanite days it was the merchants.

It was neighbours of the Canaanites, the Philistines, the people who gave their name to Palestine and who were another small trading nation, who around 1200 BC first discovered and then worked iron, so introducing the Iron Age – no great empire ever made such a pivotal discovery. Metallurgy has been one of humanity's great successes, and iron was the flourish in our exploitation of metal, inspiring early tribute:

> Hebe prepared her chariot by fixing the two bronze wheels on the ends
> of the iron axles. The felloes of the wheels are made in imperishable
> gold, with bronze tyres on the rims – wonderful work – while the
> rotating naves are of silver. The car has a platform of gold and silver
> straps, tightly interlaced, with a silver shaft to the front of which Hebe
> tied a beautiful golden yoke and fine gold breast-straps.
>
> Homer, *Iliad* , translated by EV Rieu, Penguin 1950

During the seventh century BC yet another small trading people, the Lydians, made the next important advance, the invention of coins. Previously the Bronze Age empires had not possessed coins because, being command economies, they had largely told people what to produce, what to consume and with whom to exchange goods, if any. Command economies can survive on bureaucracy alone, but trade of any volume requires money.

Barter cannot serve a brisk market. The simplest markets, those of Stone Age times, could manage by barter, but once business has become brisk, barter is limited by convertibility, or rather its lack. Barter allows traders to swap a pig for, say, twenty ducks, or three banana trees for half a deerskin, but if one pig is worth twenty ducks, and if six ducks are worth one banana tree, and if three banana trees are worth half a deerskin, then how many ducks does a trader get for a deerskin, and how many pigs for a banana tree?

The more goods that are bartered, the more exchange rates are therefore needed. Swapping three goods demands only three exchange rates, but ten goods require forty-five rates, and one hundred require 4,950. The formula for the number of combinations is, of course, $C^n_r = n!/[(n-r)!r!]$ where n is the number of commodities and r = bilateral pairs. Since the number of exchange rates grows exponentially, a common currency is required within even modest markets.

For millennia, as is revealed by the dictionary, cattle were money. Our words capital and chattel both derive from cattle, just as pecuniary derives from *pecus*, the Latin for cattle. But if two traders wish to swap a pig for twenty ducks, cattle provide a poor means of exchange. That was why one of the most successful of the early currencies was the cowrie. This little shell, the product of a mollusc that lives in the Pacific and Indian Oceans, has been the most successful currency in human history, both in geography – serving much of Asia and Africa – and in longevity. Dating from the New Stone Age and already old by the dawning of the Bronze Age (the Chinese character for money is a mollusc), it was still in use within living memory: as late as the twentieth century the British colonial authorities in Nigeria accepted cowries as payment of taxes.

But not even cowries provide a perfect currency. Molluscs reproduce, so if the shells can be harvested too easily, or if they can be disseminated too readily, a cowrie-based currency will inflate. It was Western technology in the form of the improved fishing boats the British introduced and the faster dissemination of the cowries by the railways the British built, that did for the little shells, particularly in Africa. In 1790 the price of a bride in Uganda was two cowries; by 1860 it had risen to 1,000.

Precious metals, however, which are called precious because they do not rust and are rare, make better currency. Here is the record of an early transaction:

> And Sarah died in . . . Hebron, in the land of Canaan, and Abraham came to weep for her. And Abraham spoke to the sons of Heth saying, 'I am a stranger with you, give me possession of a burying place with you that I may bury my dead . . . Give me the Machpelah cave for as much money as it is worth' . . . And Ephron the Hittite, who dwelt amongst the children of Heth, answered: 'The land is worth four hundred shekels [weighed pieces] of silver'. And Abraham weighed to Ephron the silver . . . And the field and the cave that is therein were made sure unto Abraham for a possession of a burying place by the sons of Hebron.
>
> Genesis 23:2–18

This account shows that by the middle of the second millennium BC weights of precious metal were being used as money. It also shows that both the Canaanites and Jews recognized private property rights – 'made sure unto Abraham for a possession' – and it also shows that the Jews, Canaanites and Hittites, from small trading nationalities all, shared a common culture of commerce. Since we also know from Exodus 20–23 that the rule of law was already then strong among the Jews, we can see that three of the bases for economic growth – a market, private property and the rule of law – flourished in the little nations outside the Bronze Age empires.

But as entrepreneurs the Jews and Canaanites were positively lethargic compared with the Lydians. The Lydians, a small trading nation on the Western side of modern-day Turkey, were so commercial that, as Herodotus complained, they sold their daughters into prostitution, though, as Exodus 21:7–8 shows, the Jews were not so backward with their own daughters, whom they would sell into concubinage. The Lydians were also the first people to open permanent retail shops. And, as both Herodotus and Aristotle reported, in 687 BC, under their immortal King Midas, the Lydians invented the first coins.

Coins can provide a better currency than do simple weights of metals because coins come stamped, often with a king's head, with a guarantee of metallic purity. The determination of metallic purity had long been a problem, and Archimedes had run naked through the streets of Syracuse shouting '*Eureka*! *Eureka*!, I've found it! I've found it!', when, in his bath, he thought of a method based on the weighing of displaced water. (This shows, incidentally, how markets inspire scientific advance: the market need arose first, and the scientific solution followed; it flouts Bacon's linear model, which would have predicted the opposite, namely that a scientist should have discovered how to determine metallic purity, which should then have inspired a king to mint coins.) And during the next century under another of their immortal kings, Croesus, the Lydians invented bimetalism or coins in both gold and silver.

The Greeks, meanwhile, were rapidly adopting Lydian-type coins. The Greeks, yet another small trading nation, so honoured trade that they centred their towns not on palaces, temples or castles but on their markets or agora. And they studied commerce: around 350 BC Lynceus published the world's first consumer guide, *How to Buy in the Market*. The Greeks were indeed a trading nation, and they needed to be: they inhabited a barren land, and, had they not fished and sailed and bartered all over the Mediterranean, they would have starved. They founded so many trading colonies on the shores of modern-day Turkey, Italy, France, Spain and Africa that Plato commented that the Greeks lived around the Mediterranean 'like frogs round a pond'.

THE INVENTION OF SCIENCE

And the Greeks invented modern maths (*mathematikos*, from the Greek *mathema*, science). The Sumerians, Egyptians and Indus Valley civilizations had developed some workable maths, but the closer we look the odder their maths appears. For example, an Egyptian who wanted to calculate one-third (⅓) of something would first calculate two-thirds (⅔) of it and then halve that. How bizarre. Yet that convolution illustrates a fundamental point: Bronze Age maths was empirical. The late New Stone Age/early Bronze Age had determined, by trial and error, certain arithmetic and geometric procedures that seemed to work, and thereafter those procedures, no matter how cumbersome, were followed by rote. Nobody knew how or why those procedures worked or how they could be simplified, nor did anyone apparently ask.

But the Greeks asked. They actually invented the idea of proof. The Greeks refused to accept a piece of maths until it had been proved to be true, and they invented different ways of proving it. As the fourth-century scholar Pappus of Alexandria explained in his *Mathematical Collection*, for their proofs Greek mathematicians employed five techniques, which went by the names of analysis, synthesis, reduction, exhaustion and reduction to absurdity. Pappus also discussed *diorismoi*, the determination of whether or not a solution can be proved.

It was Euclid who, around 300 BC, published his *Stoicheia (Elements)* to capture the best of Greek inquiry and scepticism. So, Euclid explained, in geometry there should be *definitions* (what is a line, what is a plane), *postulates* (unprovable assertions, such that one can draw a line between two points or that one can draw a circle around a point or that parallel lines do not intersect), *axioms* (postulates that are common to all maths, such that if a = b and b = c then a = c), and finally *propositions* ('in a right-angled triangle, the square of the hypotenuse is equal to the sum of the squares of the other two sides'). Propositions can be logically proved if the definitions, postulates and axioms are accepted.

It is only with proof, definitions, postulates, axioms and propositions that maths becomes a science, so, not surprisingly, it was also the Greeks who invented modern science. Science, Bacon said, is a matter of observation, induction, deduction and experimentation, and, as the surviving documents reveal, the Greeks mastered all four. During the fourth century BC Aristotle wrote a series of books brimming with observations on physics, biology and other sciences, and in his *Posterior Analytics* he defined induction as the generation of general principles out of particular observations: 'We have produced, from many notions gained by experience, one universal judgement about a class of objects.' Euclid had already defined deduction as the determination of particular truths by the logical application of general principles; and in *On the Method* Archimedes showed during the third century bc that the Greeks could be mean experimenters.

Observation, hypothesis, logical deduction and experimentation – the scientific method in short – suddenly arose in Greece during the sixth and fifth centuries BC. Why there, why then? In his classic *Greek Science in Antiquity* (1955) Marshall Clagett of the University of Wisconsin wrote of the 'Greek Miracle', but I think we can do better than that because on page 25 of his book Professor Clagett also wrote that the Greeks 'often exhibited an almost intuitive understanding of the proper relation of a scientific theory to observed data'.

That is the key; the intuitive understanding of the scientific method. Just as Adam Smith maintained that trade was innate, so the psychologists have shown that the scientific method is too. It was the Swiss psychologist Jean Piaget (1896–1980), who showed in his 1941 book *La Gènese du nombre chez l'enfant* how children are naturally creative and naturally logical – they are intuitive scientists. Piaget made that discovery when, working as a school teacher, he noticed that children of particular ages made consistent mistakes in logic – that is, children possess their own innate reasoning and creativity, which change as they mature into adults but which are not acquired from imprinting adult thought processes onto a *tabula rasa* but which are innate, hence the consistent age-related mistakes.

In his 1996 book *The Prehistory of the Mind: A Search for the Origins of Art, Religion and Science* Steven Mithen, the Reading University archaeologist, even showed that certain particular sciences are innate. We instinctively understand, for example, some of the mechanics of movement: babies anticipate (as shown by their eye movements) that one object, on hitting another, will cause it to move: $f = ma$ is hardwired.

So intuitive is the scientific method, indeed, that even animals demonstrate it. When dolphins that have been rewarded for touching objects of particular shape/colour combinations are presented with novel shape/colour combinations, they will successfully predict the ones that will yield further rewards. Dolphins will observe, induce and deduce.

And chimpanzees can experiment: when shown bananas that are too high to reach, they will place different combinations of boxes on top of each other until they have constructed a satisfactory 'ladder'. Going further down the evolutionary tree, my own PhD students always arrive in the lab already understanding the scientific method. When you consider the poor state of education in Britain today and when you consider the ignorance my students have of anything other than football, their understanding of the scientific method can only have been intuitive.

Intelligence is intuitive because it evolved to be adaptive – to help individuals adapt to reality. The woman who sees three lions pass behind a bush and only two emerge the other side, and who then chooses to picnic by that bush, may not produce many offspring. The man who thinks, 'Girls smile at men they like, Sue is smiling at me, I'd better kill her,' may not produce many offspring either. Intelligence evolved to match external and social realities, and if the physical and sexual laws of our universe had been different, so too would have been our inherited laws of logic.

We can now begin to understand the Greek achievement: at one level it was simply negative and trivial, namely to ignore the theocratic oppressions of the Bronze Age. People are born naturally scientific but under the oppressions of the Bronze Age they had suppressed their animal spirits of inquiry and been tyrannized into subscribing to nonsenses such that the emperor was a god. Solely with the Greeks, who had emerged directly out of the tribal New Stone Age, did humanity finally provide an urban civilization that had retained its freedoms. It was not surprising, therefore, that it was a Greek, Hippocrates of Cos, the Father of Medicine, who during the fourth century BC wrote in his *On the Sacred Disease*:

I am about to describe the disease called 'sacred' [epilepsy]. It is not in my opinion any more divine or sacred than other diseases but has a natural cause and its supposed divine origin is due to men's inexperience and to their wonder at its peculiar character.

Such naturally adaptive thoughts would have been almost impossible to conceive and, if expressed, immediately punished as heresy by the theocratic tyrannies of the Bronze Age, but they were instinctive to many Greeks.

Yet science can impose its own tyrannies, and, having introduced science, the Greeks also introduced its flaws. Consider the story of Pythagoras (of the Theorem) and of Hippasus, his student. Pythagoras was a good scientist, but he was of a mystical bent and he revered 'rational' numbers (whole numbers or whole fractions). He believed they explained the Harmony of the Spheres: Pythagoras believed that whole numbers underpinned the universe, from music to the movement of the planets.

Unfortunately, Hippasus discovered that the square root of 2, $\sqrt{2}$, is not a rational number. It is, in fact, an 'irrational' number, and its exact quantity will never be precisely calculated because, as Hippasus showed 2,500 years ago, irrational numbers can never be definitively calculated. This proof upset Pythagoras and he asked Hippasus to retract it. But Hippasus refused, so Pythagoras had him drowned.

That crime was committed during the fifth century BC, so the contemporary evidence is lost and we have to accept it on the hearsay of later Greek writers, but the fact of the story, and its credibility to those of us who know scientists (one of my research supervisors hated all his competitors and would have murdered them all), speaks of the perennial nature of the scientific personality.

The scientific personality is perennial because, as Bacon noted, the problem of the ownership of knowledge appears to be perennial. An artist's work is unique, and a householder's house is his under the law, but once a scientist has published, the work apparently becomes public property. A scientist, therefore, must strive for public esteem, because, in the absence of ownership, that seems to be the only currency of science. Yet a scientist will achieve esteem only by publishing first and by being correct – so being scooped or disproved is a disaster, and one that might justify extreme measures.

HOW DID THE GREEKS ESCAPE FROM THE BRONZE AGE?

The Greeks, like the other innovatory nations of the day, were commercial and seagoing. The two were linked because carriage by sea, even today, is ten to twenty times cheaper in terms of cost per mile than is carriage overland, so seagoing fosters trade. Seagoing is also less easily policed by a tyrant's thugs, so seagoing fosters cultural freedom: a sailor, hopping from port to port, would grow sceptical of the claims of total but contradictory certitude he would receive from the priests of different lands. The Greeks grew highly sceptical, and they evolved a light-hearted and entertaining religion that rarely imposed on people.

There were some rare exceptions. During the fifth century BC Anaxagoras taught that the stars might not be gods but 'clods of earth', that the sun might be a rock 'bigger even than the whole of the Peloponnese' and that the moon shone by reflected light. The Athenians were so shocked by his impiety that they would have executed him had not his old pupil, Pericles, arranged for his hurried escape to Lampascus. And the Athenians did actually condemn Socrates to death for sacrilege, though they probably intended that he, too, should escape into exile.

But Anaxagoras and Socrates were unlucky in their timing. It was only to enlist the support of the gods for their upcoming war against Sparta that the Athenians had passed their law against blasphemy in 432 BC. And it was a Greek, Epicurus (the hedonist who gave us epicurian) who during the third century BC attacked religion 'for producing criminal and impious deeds'. And it was Epicurus who proposed a secular law for the conservation of matter – 'no thing is ever produced from nothing by divine power' – and who elevated personal observation over revelation by faith or authority: 'What can we find more certain than the senses themselves to mark the true and the false?'

Science thrives on freedom, which in turn thrives on egalitarianism and democracy, and the Greeks were democrats. But Greek democracy was only the retention of, or reversion to, ancient practice. The Greeks didn't invent democracy – that was common to many European tribes in classical times. In *Germania*, for example, Tacitus, the Roman historian of the first century AD, described many of the German tribes as being democratic, egalitarian and free. Indeed, some German tribes even treated women as equals, though the Sitones went too far – 'among the Sitones tribes woman is the ruling sex', which Tacitus condemned as 'below decent slavery'.

Paradoxically, it was probably because Greece was barren that the Greeks retained their democratic heritage. Just as the finest wines grow on the poorest soils, so egalitarianism can flourish when times are hard. A pharaoh, presiding over the easy wealth of the Nile's annual flood, could afford to oppress whomsoever he wished; and his peasants could afford to be oppressed, for they generated an economic surplus. But a subsistence farmer stuck on the poor land of the Peloponnese could not afford to be oppressed because he generated no economic surplus; he needed everything he produced.

Egalitarianism, in recognizing people's human rights, extends that recognition to their possessions, and with secure property rights come markets and investment and therefore new technology, because the investor who designs a better mousetrap can make a profit. And, by profits, entrepreneurs create the capital for further investment.

So markets, investment and technology emerge out of a political and legal settlement that recognizes human rights, which the Greeks did. As Professor Fred Miller showed in *Nature, Justice and Rights in Aristotle's Politics* (1997), during the fourth century BC Aristotle elaborated the strongest statement of human and property rights the world had yet seen. Socrates had earlier argued for the common ownership of property, women and slaves, but in his *Politics* Aristotle linked the Athenian tradition of private property to its commercial success.

GEOGRAPHY

Greece was the gift of the Mediterranean. Small enough to allow ships to cross, the Mediterranean proved a moat big enough to help deter armies and empires. By being both large yet circular, the Mediterranean thus acted as an interface between separated cultures, encouraging a diversity of thought.

It was a close run thing. As recently as 5.8 million years ago, as Kenneth Hsü explained in his 1983 *The Mediterranean was a Desert*, the Mediterranean was dry. A dry Mediterranean might have done no more for Greece's development than the dry Sahara has done for tropical Africa's, but fortunately continents shift. Thanks to continental drift the Atlantic broke through the Strait of Gibraltar to produce: 'the Gibraltar Falls [which] were one hundred times bigger than the Victoria Falls and a thousand times grander than Niagara'. For centuries the waters rushed in at over 100 knots, eventually to generate the inland sea that, by fostering trade across the empires it kept apart, was to play such a vast part in promoting freedom and technological development.

And the Greeks were lucky in their mountains. Plains allow a king to conquer without obstruction and thus coerce a vast population under his monolithic rule, but mountains are hard to scale and easy to defend. Greece, like much of Europe, is mountainous, allowing its different peoples to develop in discrete ways, thus promoting a diversity of thought.

WHY DID SCIENCE EMERGE IN MILETUS?

Remarkably, the world is agreed on who was its first scientist, Thales of Miletus. His exact dates are not known, but he was alive in 585 BC, successfully to predict that year's solar eclipse. Scientists number two and three were, respectively, Anaximander of Miletus (c.610–545 BC) and Anaximenes of Miletus (died around 500 BC). Miletus was, clearly, the birthplace of science, and science starts with Thales, Anaximander and Anaximenes in Miletus during the sixth century BC. Why then, why there?

Here is a clue: Miletus was the most commercial of the Greek cities. Banged right up against Lydia – and we know how commercial that was – it was the biggest and most prosperous trading city of its day. Thales, Anaximander and Anaximenes, moreover, were businessmen, and businessmen often travel, which broadens the mind. Here is the fifth-century philosopher Proclus:

> Geometry was discovered by the Egyptians to help them resolve land disputes after the Nile had flooded each person's land . . . It was Thales who, after a visit to Egypt, first brought this study to Greece.[1]

If Thales had done no more, he would still have been a colossus – he copied Egyptian geometry! That might now seem trivial, but the empires of the Bronze Age did not copy. Life in the imperium was predetermined, people did what they were told, and someone who copied was either copying a different caste – for which the penalty, under the sumptuary and other laws, was death – or copying a foreigner, for which the penalty was death.

Copying comes naturally to children, it is instinctive. Copying, being an early step in learning, is also a basis of scientific, technological and economic development. Academics call learning by copying 'reading the literature', engineers call it 'technology transfer', and businessmen call it 'following best practice'. Copying is natural to markets. Where would Burger King or Wendy's be today if they hadn't copied McDonalds's? Without learning by copying, every individual has to reinvent the wheel.

But Thales was not solely a copycat, as Proclus also said: 'Thales made numerous discoveries himself, attacking some problems systematically and others empirically.' So, how did Thales conceive of actually doing science? Let us look to the famous story Aristotle tells in his *Politics* of Thales and his poverty:

> When challenged as to why he was so poor, Thales replied that he had once determined by astronomical observation that the olive crop would be good that year so, it still being winter, he raised some capital and bought the rights to all the presses. Later that year he rented out those presses at a good profit, thus proving that philosophers can make money if they choose to, but they often choose not to.[2]

Consider what the story reveals about the market and scientific methods. First, consider what Thales, as a merchant, does. Initially, Thales the merchant makes an observation: the stars have attained a certain conjunction. Next he creates an hypothesis: the conjunction will cause the olive trees to yield a good crop. Next he tests his hypothesis: he buys the rights to the presses. Finally, he measures the outcome of his investment: he makes a good profit.

Now consider what Thales, as a scientist, does. First, Thales the scientist makes an observation: the stars have attained a certain conjunction. Next he creates an hypothesis: the conjunction will cause olive trees to yield a good crop. Next he tests his hypothesis by experiment: he buys the rights to the presses. Finally, he measures the outcome of his experiment: he counts the profit.

Do you get it? The scientific method is only the market method; the market method came first, historically, and the scientific method emerged when a trader, Thales, first extended his market method into an abstract problem of the type we call scientific. What the directors of a company or the dealers on the exchanges or the entrepreneurs in the market do today is no different, in kind, from what researchers do in their laboratories, but it was the traders who taught the scientists how to formalize it.

Consider, moreover, the scientist's obsession with proof. A Bronze Age bureaucrat, secure in his unchallengeable control of the proles, need never have worried about the principles of his managerial or technical procedures. As long as they worked tolerably, who cared? The trader cared. His life was uncertain. Thales was not the only Miletan capable of making astronomical observations. A competitor who could make better observations might do better in the olive press futures market and so drive Thales into bankruptcy. Thales would *really* have wanted to know how, and why, exactly, the stars regulated the olive crop. Investors ache for the greatest possible certainty before handing over their money. They therefore create a culture where scientists will seek to prove their assertions.

* * *

SEX, SCIENCE AND PROFITS

The scientific method emerged in Greece, therefore, because Greece was a land of markets; and the market method, which is the application of our instinctive logic and creativity to the optimization of wealth, is rational because markets are real – as real as Sue's smile or the lion that lurks behind the bush. The Bronze Age, by crushing reason under the weight of emperor worship, had destroyed humanity's recourse to personal instinct but the Greeks, vulnerable to the vicissitudes of the elements, had sustained reason.

Thus did Aristotle claim Thales as the world's first scientist. Today some people may suspect the idea of a 'first' scientist – some scholars deny that Thales predicted the eclipse of 585 BC – but regardless of the details, Thales was revered by the Greeks because in a theocratic world he was the first to invoke reason in explaining Nature.

And then, having revered reason, the Greeks lost it.

5
The First Baconian Experiment

'History . . . is, indeed, little more than the register of the crimes,
follies and misfortunes of mankind.'
Edward Gibbon, *The Decline and Fall of the Roman Empire*,
1776–88

The Greeks flourished by trade and by technological development,
but then they did an odd thing: they largely abandoned them.
Consider the social status of merchants. Once it was high but, only two
centuries after men like Thales of Miletus were being lauded,
fashionable Greeks were openly despising the *banausoi*, the men who,
as bankers, traders or doctors, actually earned their living. In his
Politics Aristotle maintained that it was poor taste even to discuss the
banausic professions, and the classic expression of contempt for the
banausoi was made by Xenophon, the Athenian-born general and
writer who commanded the 10,000 Greek soldiers in Persia and who
had once been a pupil of Socrates:

> The banausic trades spoil the bodies of workmen and foremen who are
> forced to sit still and work indoors. They often spend the whole day at
> the fire. The debilitation of the body is often accompanied by a serious
> weakening of the mind. Moreover, the banausic professions leave no
> spare time for service to one's friends or the city. Thus the *banausoi* are
> considered unreliable friends and poor defenders of their country.
>
> *Oeconomicus*, IV, 2–3, Harvard University Press, 1923

Why were Aristotle, Xenophon and their upper-class friends so snobbish about the trade and the technology it had spawned that had enriched Greece? The answer is that Aristotle, Xenophon *et alia* had learned that warfare, not commerce, were emerging as surer routes to personal wealth. They were learning the lesson of empire: that in a politically insecure world, in which farmers are tied to the soil and artisans to their workshops, the predator can out-earn the merchant. The market may have provided the Greeks with a technical fillip, but war subsequently enriched them more, and war substituted aristocratic values for banausic ones.

So Xenophon became a well-paid mercenary for the Persians, the Greeks' traditional enemies; and then he fought for the Spartans, the Athenians' traditional enemies. Those successive betrayals drove his own people, the Athenians, to banish him – and Xenophon had claimed the *banausoi* were 'poor defenders of their country'! But what did he care? He ended his days with a grand estate at Scillus; few merchants could equal that. Aristotle, meanwhile, had watched King Philip II of Macedonia conquer all of Greece, and he had watched his own pupil, Alexander the Great, conquer half the known world – the Persian Empire. And he had noted that both monarchs, father and son, had funded his own academic research programmes with a generosity no market would, or should, have done. To Xenophon and Aristotle, therefore, it seemed obvious that imperialism represented the surest route to country estates and to research grants.

Still, the Greeks had once been creative; the Romans were plunderers from the beginning.

ROME

Edward Gibbon's *Decline and Fall of the Roman Empire* ran with a question that people had been asking ever since the Roman Empire did decline and fall: why did it? Some people have blamed lead in the water, others the endless civil wars and the lack of proper democracy, and yet others the army for its indiscipline and overdependence on barbarian recruits. Gibbon himself believed that its conversion to Christianity had infected the empire with moral decay and superstition – 'I have described the triumph of barbarism and religion' – but Gibbon, who twice converted between denominations, was unsound on faith.

To an economist the fall of Rome is straightforward. It can be explained by Mr Micawber: 'Annual income £20, annual expenditure £19, 19s, 6d, result happiness. Annual income £20, annual expenditure £20, 0s, 6d, result misery.' Rome was a plunder empire. It started that way in 753 BC, raping the Sabines' women and stealing their cattle, and, as it expanded, so it continued to finance itself by feeding off its victims. The Romans butchered their way away across the Mediterranean. Theirs was, admittedly, intelligent butchery – Rome was clever at recruiting its victims to its cause – but Rome's achievement was essentially military and political.

But when, at its maximum expansion, the empire ran out of lucrative victims, its expenditure exceeded its income, and the result was misery. Because the Romans were not technically inventive (they invented little other than the grotesqueries of the Colosseum), they had few alternative sources of income. The Roman Empire, particularly in the West, basically went bankrupt.

The two great scholars of technology in antiquity, Marc Bloch and Moses Finley, scorned the Romans as luddites. Consider the water-mill. This machine, by harnessing the power of water, eased and cheapened the grinding of corn. It had not, of course, been invented by the Romans – it was a barbarian development (from Jutland, around the time of Christ) – yet, as Bloch and Finley complained, the Romans failed even to adopt it:

The astonishing thing about the watermill is that, having it at their disposal, the Romans were so slow at bringing it into general use.

Translated from Marc Bloch, *Avènement et conquête du moulin à eau*, 1935

Every rational argument suggests the quick and widespread adoption of the watermill, yet though it was invented in the first century BC, it is not until the third century AD that we find evidence of much use.

Moses Finley, *Technical Innovation and Economic Progress in the Ancient World*, 1965

Scholars today are slightly kinder to the Romans. Bloch and Finley depended largely on written records, yet the literary elite of Rome rarely deigned to notice machines, but as new archaeological evidence emerges so, correspondingly, the Romans emerge as more open to new technology. Thus it transpires that the Romans did install watermills – quite a number, actually – and that they even improved on the barbarian designs, converting the mills from horizontal to vertical flow.[1] And at least one poet, Antipater of Thessalonica, who wrote during the time of Christ, celebrated them:-

> You can sleep, milk maids
> And rest your tired bones
> It's now the water nymphs
> Who spin those mill stones

The Romans made a few other technical advances. Around 200 BC they improved on the concrete of the early Egyptians, developing a variety that would set under water. Coupled with their enrichment of vaulting techniques, this enabled them to erect some dramatic buildings (though the famous round arch had been an Etruscan, not

Roman, innovation). To coat their swords with steel, the Romans discovered how to carbonize iron. They also invented some powerful siege weapons and developed a harvesting machine for corn.

Yet that about summarizes Rome's technological developments. It is a meagre record for an empire that once dominated almost all its known world.

TECHNOLOGY AND THE EVOLUTION OF ECONOMIES

The trouble was that Rome was not a commercial empire – it knew how to raise GDP per capita only by plunder. The Romans had inherited markets and money of course, and they knew how to use them, as is revealed by some letters, written on slivers of paper-like wood, that have been excavated from Hadrian's Wall. They are dated to around AD 120. Here is an extract:

> I have bought 5000 *modii* of ears of grain, for which I need cash. Unless you send me at least 500 *denarii* I shall lose the deposit. Please ask Tertius about the 8½ *denarii* which he received from Fatalis and which he has yet to credit to my account.
>
> Quoted by A.K. Bowman *et al.*, 'Two letters from Vindolanda', in *Britannia*, 1990, 21:36–52

Rome therefore was a moneterized society but not intensively so: its level of commercialization was not enough to generate new technology. Moreover, it failed to create the corporation or firm. As Adam Smith showed, specialization is the basis of economic growth, and the classical world had created specialists. Consider the Greek specialists: the *naukleros* traded his own ship, the *emperos* subcontracted the passage of his goods to another's ship, the *kapelos* traded from home.

To support such trade, Greek law elaborated a commercial code, recognizing property rights and enforcing private contracts. Early banking emerged. Within their own empire, the Romans had evolved a similar degree of specialization. But once the benefits of individual specialization have been exhausted, market economies need to rise to the next level of organization, to that of the firm. Neither the Greeks nor the Romans made it.

FIRMS AND THE PROBLEM OF SPECIALIZATION

Specialization creates problems because specialists are not self-sufficient, they need to trade their goods with each other. But trade is difficult. When a person buys a product, how does that person know that the product is good? *Caveat emptor* – traders need to check each others' credentials and products. And contracts need to be policed by accountants and lawyers. These checks and policings consume both time and money, and the costs are known as 'transaction costs'.

It was the economist Ronald Coase who, in a famous 1937 paper 'The Nature of the Firm',[2] showed that when an economic enterprise is big enough to demand the cooperative efforts of many people, it makes better sense for them to work together as employees of one firm rather than to each operate as individual contractors. The transaction costs of every individual bargaining with every other individual over their services and contracts would be huge, but those individuals' joint employment as employees of one company provides for cheaper cooperation.

We in Britain have recently been reminded of this wisdom of Coase's. When John Major's government during the early 1990s decided to privatize the railways, it did not choose the obvious model by which the railway companies ran trains they owned on track they owned. Instead, Major separated the ownership of the track from the companies running the trains, while the track-owning company in its turn subcontracted out the actual laying, maintenance and inspection of the track to myriad subcontractors. Major hoped to increase competition, but he succeeded only in transforming the railways into a morass of contracts between contractors, subcontractors and subcontractors' subcontractors. Within a decade that morass had generated journey times and safety records that some commentators have compared unfavourably with those of the 1860s. The railways were being crippled by an excess of contracts. Unitary firms would have done better.

But firms themselves do not come cheap. Firms incur costs of their own, including those of training, supervision, management and coordination: employees need to be monitored. Consequently, the creation of firms is not to be undertaken unadvisedly, lightly or wantonly; it is to be undertaken only when the market is so thick that its transaction costs justify their creation, which the Greek and Roman markets were not. So neither the Greeks nor the Romans developed joint stock companies, nor business capital independent of private wealth, nor banks bigger than pawnshops. And they certainly did little research into new commercially useful technology.

Research into new technology is expensive and risky. Rational people will not pay for it until they have exhausted the benefits both of specialization and of company formation. But the only companies the Romans created were, ominously, companies of *publicanii* or tax farmers. Since the Greeks and Romans never felt sufficiently pressurized to form commercial companies, we can see that their economies did not demand research into new technology – they hadn't exhausted the capacities of the technology they already possessed. Why invent a new watermill when there is no demand for more or cheaper milling?

War was a major drag on the Roman economy. Consider the shrine of Janus. Janus was the Roman god of doorways, and he was represented as facing two ways – hence the month of January, which faces both the old and new year. Traditionally, the doors of Janus's shrine in Rome were closed during times of peace but, according to Livy, the doors were closed only twice during the 600 or so years between the seventh and first centuries BC. Between 200 BC and the time of Christ, moreover, on average no fewer than 100,000 men – some 13 per cent of the male citizenry – were in arms each year.

A war-disrupted society provides a thin market because entrepreneurs are not faced with intense competition from other entrepreneurs; they are faced with the problems of their sector's survival. The entrepreneur whose customers have been called to arms does not find himself pressured into doing research to improve his technologies; instead, he finds himself pressured into surviving. Moreover, if his competitors are called up, he might find himself enjoying monopoly windfalls – and those do not foster research either.

So uncommercial did the republic become that in 342 BC the Senate passed the *lex Genucia* to prohibit usury or the payment of interest on loans. This law, which was regularly renewed, naturally deterred lending, which in turn inhibited economic development. Indeed, in 218 BC the Senate passed the *lex Claudia* to prohibit senators from engaging in commerce: their job was to prepare for war and to run their country estates, not to engage in vulgar trade.

The Romans actually resisted innovation. During the first century BC some Syrians had discovered how to blow glass – a nice advance – yet, as Petronius explained in his *Satyricon*: 'A flexible glass was invented, but the workshop of the inventor was completely destroyed by the Emperor Tiberius for fear that copper, silver and gold would lose their value.' Equally, Suetonius described how: 'An engineer devised a new machine which could haul large pillars at little expense, however the Emperor Vespasian rejected the invention, asking "Who will take care of my poor?"' No wonder the Romans produced so little useful technology.

THE FIRST BACONIAN EXPERIMENT

Not all the rulers in classical times resisted innovation. Francis Bacon wrote *New Atlantis* in 1610 (published 1627) to describe a scientific utopia, his New Atlantis being a fictional island state ruled by a community of scientists. His model for New Atlantis was Egypt under the Ptolemies.

In 323 BC, on Alexander the Great's death, one of his generals called Ptolemy acquired Egypt, and until 31 BC the country was ruled by his successors, all of whom, if male, were also called Ptolemy and, if female, Cleopatra. Following Egyptian custom, royal brothers and sisters married, so the famous Cleopatra whose affairs with Caesar and Mark Anthony eventually cost Egypt its autonomy was actually Cleopatra VII, the wife of her brother Ptolemy XIII.

During their 300 years in power the Ptolomies followed the policy that Bacon was to admire: lots of government-funded science. Scientists were supported generously, not just with salaries but also with the famous library and experimental museum (so-called from the Muses, the goddesses of poetry, music, drama and so on). And the scientists were superb: they included Euclid of the *Elements*, Eratosthenes, who correctly calculated the diameter of the earth to within 50 miles, Archimedes of *Eureka!*, Ctesibius, who discovered the elasticity of air, Hero, who developed the *aoelipe* or the world's first steam engine, and Ptolemy the astronomer and geographer, whose maps and writings were still being studied in the time of Henry the Navigator.

The Ptolomies' tax-funded science rebounded, as they intended, to their international prestige. Philo of Byzantium wrote: 'Success in this work was recently achieved by the Alexandrian engineers, who were heavily subsidized by kings eager for fame and interested in the arts.' And the Ptolomies' researchers also advanced military technologies: Archimedes developed catapults, ship-levering hoists and burning mirrors, and Dionysius invented a machine gun for arrows. Their researchers also developed religious technologies because the Ptolomies, as good emperors, supported piety. They were delighted when Hero of Alexandria invented hydraulic pumps that opened and closed temple doors as if by magic!

And the Ptolomies did everything they could to encourage economic growth – everything, that is, except for promoting property rights, the rule of law or a free market. Consider money. Er, what money? As Karl-Heinz Priese showed in his *Giro Systems in Hellenistic Egypt* (1910), the Ptolomies provided the Egyptian economy with banks but not with money (coins might empower the proles). Instead, the Ptolomies provided a complex of giro banks to allow individuals to transfer payments from one account to another in notional weights of grain, which also allowed the state to monitor and tax every transaction.

Ptolomeic science did not translate into commercially useful technology because, in the absence of a market, Egypt lacked the incentives or mechanisms by which science could be applied commercially. Even Aristotle had noted the fruitlessness of the academic science of his day: 'The geometricians cannot apply their science. When it comes to dividing a piece of land . . . the mathematicians know theoretically how to do it, but not in practice.'[3] But the application of science requires commerce, and commercial values were despised. When Euclid was asked by a student for a reward for solving a problem, he tossed the student a drachma sneering, 'he wants science to be profitable'. Yet the Alexandrian scholars did not disdain their own salaries: the physicist and inventor Ctesibius, on being asked what he had gained from his researches, replied 'free dinners'.

Such a culture prevented the Ptolomies' Baconian science from translating into economic growth, thus showing that Bacon's academic science was not a sufficient nor perhaps even a necessary prescription for useful technology – Smith was right: only markets generate useful technology.

THE FALL OF ROME

Because Rome was not profitable without conquests, its decline was inevitable – and frightful. It was at the height of the empire's expansion, at the beginning of the second century AD, that the underlying economics kicked in. By then the costs of supplying the garrisons was horrendous but, having run out of fresh booty, the state could support them only by raising endless taxes.

To make sure that people paid the taxes, their freedoms were increasingly taken away, and armies of secret policemen were recruited as tax collectors. As the economy collapsed under their weight, Diocletian (reigned 284–305) eventually imposed fixed prices and wages, with infractions punishable by death. What was left of the market was so crushed that by the third century AD most banking, which had never been extensive, simply disappeared. Property rights and the rule of law were obliterated as the emperors simply plundered the rich. And in 332 Constantine bound all tenant farmers, known as *coloni*, to the state as serfs – their children were *glebe adscripti*, tied to the soil – while the city guilds or *collegia* imposed compulsory, hereditary trades on their workers: an edict of 380 forbade the children of the workers in the mint from marrying outside their caste or trade. Inflation, meanwhile, set in. In the mid-first century AD the silver content of the sestertius was 97 per cent, but it was 40 per cent by 250, and 4 per cent by 270.

Contrary to myth, the empire did not collapse in the face of unstoppable barbarian hordes. The numbers of barbarians were often small – a mere 80,000 Vandals took the whole of Roman Africa in less than a decade. The empire fell because its citizens despaired. Its population had already fallen from 70 to 50 million as people ceased to breed or emigrated to barbarian lands. Not, indeed, that the barbarians were that barbarous. After centuries of trade with Rome they understood commerce, they coined money, many of them were Christians, and they were often literate. And they were generally welcomed by the common folk as armies of liberation.

The fall of the Roman and Ptolomeic empires thus teaches that when all the entrepreneurs in a society are soldiers or politicians, neither strong government nor academic science will save it. This point is too rarely understood. A standard text such as R.A. Buchanan's *Technology and Social Progress*[4] teaches, in the author's own italics, on the very second page, that '*a strong state, in short, is a necessary precondition of industrialization*', but, historically, the opposite has been true. In the past it was the strong states that suppressed technology, and the weak ones that fostered it, because the weak ones were too weak to rob individuals of their economic freedoms.

To confirm the universality of the European experience, and to show that ultimately there was nothing special about Europe, let us examine the similar history of China.

CHINA

Francis Bacon may have described gunpowder, the magnetic compass and printing as the magic trio that had 'changed the whole face and state of things throughout the world', but Bacon really meant that they had changed the whole face and state of things throughout Europe, because China had, long before the Europeans, possessed the magic trio. The Chinese had also, long before the Europeans, developed efficient harnesses for horses, paper rolling, mechanical clocks and even wheelbarrows. But after three of the most exciting New Stone, Bronze and Iron Ages on the planet, the Chinese Empire stagnated, and the Portuguese were only the first of a succession of long-nosed white devils to sequester parts of it.

The story of Chinese petrification is conventionally told as one of two dynasties, the Song (960–1279) and the Ming (1368–1644). The Song are the 'goodies', overseeing exciting developments – Pi-Sheng, for example, invented movable type in 1041, 400 years before Gutenberg – and the Ming are the 'baddies', overseeing stagnation. But both, actually, were baddies.

For centuries after the time of Christ, China had been ruled centrally and powerfully, so it had stagnated. But the dynasty of the day, the T'ang, was eventually to suffer defeat after defeat until, in 755, it was undermined by An Lushan's rebellion. Although the T'angs did not immediately abandon the throne, they effectively ceased to govern. They stopped collecting whole categories of taxes, they surrendered their ownership of the land, and they shrank the army. So weakened was the state that between 907 and 959 there were no fewer than five dynasties in China, and the nation was split into ten kingdoms. This was chaos and everybody deplored it.

And the consequence was vast increases in wealth and vast developments in technology. Consider the production of iron. Centuries before the Europeans, the Chinese possessed blast furnaces and were using coke. Here are some production figures:[5]

Year	Annual iron production in tons
806	13,500
998	32,500
1064	90,400
1078	125,000

This was exponential growth! And it occurred 700 years before Britain was to experience the same. Indeed, as late as 1788 Britain was still producing only 76,000 tons of iron annually. But China was humming. Between 1024 and 1107 its merchants introduced paper money – 500 years earlier than in the West – and thereafter invention after invention tumbled forth. Meanwhile, their ships roamed the China Sea and inland rivers in search of every possible trade with peoples as diverse as the Indonesians, Koreans and Taiwanese.

And what was the response of officialdom? Grave disquiet. As the writer Ting-ch'iao bemoaned during the 1330s: 'Wherever there is a settlement of ten households, there is always a market for rice and salt.' (Quoted by William McNeill, *The Pursuit of Power*). A market for rice and salt! Officialdom was outraged. The distribution of rice and salt was properly a government monopoly but, thanks to market incentives, people were building canals, transporting foodstuffs, breeding new strains of crops, developing new systems of irrigation and inventing new systems of fertilization. China was going to the dogs. In 983 an individual called Chiao Wei-Yo was even impertinent enough to install, on the Grand Canal near Huai-yin, the world's first lock gates.

China had been warned. During the 1040s the senior mandarin Hsia Sung had lamented:

> Since the unification of the empire, control over the merchants has not
> yet been well established. They enjoy a luxurious way of life, living on
> dainty foods and delicious rice and meat, owning handsome houses
> and many carts, adorning their wives and children with pearls and
> jades, and dressing their slaves in white silk.
>
> Quoted by William McNeill, *The Pursuit of Power*, Blackwell 1983

A luxurious way of life! Horror upon horrors! The merchants had
emerged because the state, under the T'angs, had become too weak to
keep them down. But after the Songs had reunited the country under
firm leadership, that lapse could be corrected. First, the state
nationalized the sale of iron and salt, then it raised taxes, then it raised
a standing army of no fewer than one million men out of a total
population of about 100 million, then it raised taxes again, and then it
nationalized the printing of banknotes.

Then, wreaking such economic havoc as to render its peoples
helpless and starving, the emperors printed so many banknotes and so
inflated the currency that, for 500 years thereafter, no notes were used
again in China. And the Chinese state obliterated property rights and the
rule of law, simply killing merchants or confiscating their money at will.

It took time, of course, for the Songs to completely crush the
creativity of the market they inherited so, paradoxically, they actually
presided over a technological and economic renaissance. But finally,
with the Mings, everything returned to pharaonic-like normality, and
China settled to centuries of stagnation. The standard technological
texts on engineering, architecture and medicine were reproduced
unchanged for centuries, and once-useful inventions such as gunpowder were put to only frivolous use: the emperors enjoyed a nice
firework display but they forbade the application of gunpowder to
anything practical, such as quarrying or firing guns.

The Chinese sailed, but not to trade. In 1405 the Ming emperor,

Yung Lo sent the Three-Jewel Eunuch Admiral Cheng Ho, in command of 37,000 men and sixty-two ships, on the first of seven missions to assert Chinese hegemony from Java in modern Indonesia to Somalia in the Horn of Africa. This was impressive navigation, but it achieved nothing concrete, nor was it meant to. The imperial court in Peking (Beijing) was too grand to trade with barbarians – it wanted only to impress them with China's superiority and self-sufficiency, so the trips were intentionally fruitless (other than the giraffe Cheng Ho brought home), and in 1433 the emperor reimposed his ban on foreign travel, decreeing ever more terrible deaths for those who spoke or traded with foreigners.

ARABIA AND INDIA

To confirm the universality of the Roman and Chinese experiences, let me briefly note the similar histories of two other cultures we have already encountered, those of the Arabs and the Indians.

The leading scientific texts available to Henry the Navigator when he embarked on his campaign of exploration were not European but Arabic, including Ibn Sina's (Avicenna's) *Canon of Medicine*, al-Haytham's (Alhazen's) works on optics and, most importantly, al-Khwarizmi's *Algebra, Arithmetic, Geography* and those books written by his followers. Islam had long been more academically advanced than Christendom, and it was to Muslim scholarship that inquiring Christians of Henry's day looked. Indeed, al-Khwarizmi was so important that his name is still remembered in the word algorithm, meaning a systematic procedure for calculation. He also gave us the word algebra from his book *Algebra*, whose full title was *Kitab al-jabr wa al-muqabalan*, the word *al-jabr* meaning restoration, the calculatory technique al-Khwarizmi used.

Of all the services al-Khwarizmi performed, though, perhaps his greatest was the introduction of so-called Arabic numerals. Previously, Mediterranean people wanting to count had used the Roman numbers I, V, X, D, C and M, which are almost impossible to manipulate mathematically, but al-Khwarizmi introduced the decimal numerals of 1, 2, 3 and so on. Yet, just as the Arabs had traded other people's spices, so al-Khwarizmi had borrowed other people's numerals, for he did not actually invent decimal numbers, he only borrowed them from India where they had been in use for two or three centuries. We should really call decimals Indian, not Arabic, numerals.

Northern India, when it was first ruled by the Gupta dynasty, was a ferment of creativity. The Guptas not only presided over the invention of decimal numbers, they also presided over the development of the sine theorem in geometry, the invention of chess, metal-working of a high order, a rich literature and trade with China and the Near and Far Easts. But from the fifth century AD the Guptas, Song-like, presided over the tightening of the caste rules that were to shackle India's dynamism.

And the Arabs, too, were to petrify. During the later part of the first millennium AD Islam had, in its preservation of Greek and Roman writings, in its transmission of Indian learning and in its own philosophers and scientists, proved more questioning, more scholarly and more receptive to innovation than had its Christian or Jewish contemporaries. Indeed, it was the Arabs who inspired the great Jewish renaissance in philosophy: it was during the ninth century AD that, inspired by their Muslim neighbours, Jewish scholars started to address the classic Greek texts, culminating in the twelfth century work of Maimonides, whose *Guide for the Perplexed* (1190) was written in Arabic.

The Muslim world was even to produce a precursor of Francis Bacon, Rashid al-Din (1247–1318). In a remarkable career that in many ways anticipated Bacon's, Rashid rose to become vizier or chief minister to the Persian Empire. A scholarly man, he collected all the knowledge open to him in his vast *Jami al-tawarikh* (1302). Rashid was the first modern writer to argue for the government funding of science, claiming, some 300 years before Bacon, that 'there is no greater service than to encourage science and scholarship', and that 'scholars should work in peace of mind without the harassments of poverty'. But by his day Islam was slipping into the usual theocratic intolerances, and Rashid was executed for blasphemy, after which his head was carried through the streets of Tabriz with the cries of: 'This is the head of the Jew who abused the name of God. May God's curse be upon him!' That language of theocratic intolerance was to presage the Persian and Arabic decline into petrification.

Consider the Islamic response to printing. It was Johann Gutenberg of Mainz, Germany, who, around 1450, invented movable type and who started to print books in large numbers. It would be wrong to suppose that the local civil and religious authorities were delighted by his invention, for they were not. They were suspicious – from their perspective they should have been – because printed books were to help stimulate the Reformation. But Gutenberg's right to print was protected by Johann Fust, his business partner, who resorted to the law and won his case. The local civil and religious authorities did not, therefore, destroy the new technology, they sought merely to regulate it within the law.

And regulate it they did. By 1486 the Frankfurt censor had banned all vernacular translations of the Bible, and soon afterwards Pope Innocent VIII ruled that all publications had to receive a bishop's licence – *nihil obstat* (there is no impediment) and *impramatur* (let it be printed). In 1557, moreover, Pope Paul VI established the Catholic Church's *Index Librorum Prohibitorum* (Index of Prohibited Books). The Index was discontinued only in 1966, at which time it contained some 4,000 titles, including Francis Bacon's books. The Protestants proved almost as suspicious of the press, and William Tyndale, to mention but one martyr to scholarship, was executed in 1536 for translating the Bible into English.

Yet these Christian censorships were as nothing to those of Islam. The Catholics may have banned 4,000 printed books, but the Muslims banned all of them, and the first printed Muslim books did not appear until the nineteenth century. As Atatürk, the founder of modern Turkey, complained on 5 November 1925, when opening the new law school in Ankara:

> Three centuries of observation and hesitation were needed, of effort and energy expended for and against, before antiquated laws and their exponents would permit the entry of printing into our country.
>
> Bernard Lewis, *The Emergence of Modern Turkey*,
> New York, 2001 (new edn), p.274

Indeed, it is striking how most of the cultural wounds of different societies have been self-inflicted. Nobody forced the Romans, Chinese, Indians or Arabs into intellectual sterility; they chose it. Of course, it was not the 'Romans', 'Chinese', 'Indians' or 'Arabs' who made those choices, it was their rulers who, in their own interests, crushed the common people and their creativity.

That is why, unexpectedly perhaps, economic historians do not laud Gutenberg as the hero of the story of printing – original ideas like Gutenberg's are not rare – what is rare is the institutional climate that allows creative ideas to flourish and that rewards wealth creation, not rapacity. For economic historians, therefore, the hero of the story of printing was the rule of law in Mainz, which allowed Gutenberg's business partner, Johann Fust, to outface the Church in court.

* * *

As recently as 1776 Adam Smith could write of China and India that each 'is a much richer country than any part of Europe', yet even then Henry the Navigator's European successors were colonizing both nations. We humans are instinctive traders, but we are also instinctive predators, and the instinct that predominates depends on the institutions in which we find ourselves. The nations of the East had crushed the market in the interests of the rulers, Bronze Age-style, but the Europeans fostered the market, and they triumphed.

6

The So-called Dark Ages

'I plant my beans below the southern hill,
The grass grows thick, the beansprouts all too few.'

Tao Qian (365-427), *Returning to Live on my Farm*

On Easter Sunday 1341 Francesco Petrarch, a celibate in minor orders, ascended the Capitol in Rome to be crowned as the first poet laureate in nearly 1,000 years. Petrarch, who was thirty-seven at the time, was already an international celebrity, and though his private life was pathetic – he exhausted his life in the unrequited worship of a woman called Laura, a happily married mother of eleven – his public life thrived as he wrote poem after poem in worship of the unattainable Laura.

Such an exquisite would always need patrons – Petrarch lived off a succession of them – and in gratitude he described his era as a Renaissance during which enlightened plutocrats fostered writers such as himself in a rebirth of classical learning. The intervening years between the fall of the Roman Empire and his own time, however, he damned in *De viris illustribus* as a Dark Age during which rude savages had roamed the land in ignorance of the simplest rules of metre or of rhyme.

The fall of the Roman Empire had indeed brought little joy to aristocrats, poets, scholars, scientists or season ticket holders at the Colosseum, but for ordinary people it proved a liberation. During the 450s, for example, the German ecclesiastic, Salvianus of Marseilles had catalogued in *De gubernatione dei* the superior economic and political freedoms of the barbarians, explaining that many Romans had emigrated: 'Seeking a Roman humanity among the barbarians because they could no longer support barbarian inhumanity among the Romans.'

And that humanity was to translate into new technology. In his *Annals* Tacitus, the first-century Roman historian, had dismissed competitors from outside the empire: 'Barbarians are ignorant of military engines and the management of sieges.' But by the fifth century in his *Historia nova* the Byzantine historian Zosimus was bewailing 'the inventiveness of the barbarians' that had allowed the Visigoths to cross the Hellespont on rafts. A century later the Byzantine historian Procopius recorded in his *Gothic Wars* that the Huns had developed a battering ram superior to those of Byzantium or Persia.

Contrary to myth, the Dark Ages witnessed the revival of technological innovation, and some of the barbarian advances were of the most profound importance. Consider the crank: there are two types of mechanical motion, rotary and reciprocal, and it takes the crank to convert one to the other. Without the crank only the crudest machines can be built, and the crank is second only to the wheel in technological significance. Yet the Romans did not have it, and its first reference in the West is an illustration in the Utrecht Psalter dated to 816–834.

Nor did the Romans have soap. To wash they first oiled themselves and then scraped off the oil with a *strigil*, a sort of spatula. Smell-wise this technology worked, because body odour is carried by the volatile fats of the sweat glands, which dissolve in oil, but soap provides a cleaner wash. In his *Natural History* Pliny the Elder (23–79) wrote that 'soap was an invention of the Gauls', and during the second century Galen, the Greek physician, reported in his *Works* that Arateus had seen the Gauls use it to wash their clothes. Arateus used it in his own bath, but F.W. Gibbs, the modern historian of soap, believes that Arateus was probably the only Roman so to do.[1] After the fall of the Empire, however, the use and manufacture of soap spread widely, and by the seventh century there were enough soap-makers in Italy to support a craft guild or *arti*.

More important than soap were horses, yet the Romans used those with bizarre incompetence. In his 1931 *L'Attelage et le cheval de selle a travers les ages*, his comprehensive study of the horse in antiquity, Richard Lefebvre des Noëttes, a retired commandant of the French cavalry, chronicled the extraordinary story. As early as 4000 BC, judging by the recovery of horses' teeth showing the wear of metal bits, horses were being ridden in the Ukraine. But it was not the Greeks nor the Romans, but the barbarians of the Asian steppes who, around the first century BC, invented the saddle.

And stirrups – though initially accommodating only the big toe – were not added to the saddle by the Romans but by the inhabitants of Afghanistan or northern Pakistan: there is a first-century Kushan engraving in the British Museum of a big toe stirrup hooked up to a saddle. Nor did the Romans develop the full-foot stirrup; that seems to have emerged in the plains of central Eurasia, first being mentioned in a Chinese document of AD 477 (the stirrups in the film *Gladiator* were a Hollywood fiction).

Horse drawing is more important than horse riding, yet the Romans also failed to exploit that efficiently. The Romans had no horseshoes, so hooves broke all the time; they harnessed their horses with a yoke, so as the horses pulled they strangled themselves; nor did the Romans know how to harness horses in front of each other, so limiting multiple haulage. But during the ninth and tenth centuries the northern Europeans solved all three problems, inventing nailed horseshoes, the horse collar (which threw the weight onto the animal's shoulders) and the tandem harness. By these inventions the Europeans extracted four times as much power out of the horse, relegating the less useful ox to minor work; and the full exploitation of horse power so facilitated the introduction of the heavy plough and the conversion from two- to three-field rotation that in some areas agricultural productivity doubled. No wonder that parts of Europe entered the second millennium AD suddenly more vital.

Wind power was another great Dark Age exploitation. As the distinguished historian Edward Kealey showed in his authoritative book *Harvesting the Air: Windmill Pioneers in 12th Century England* (1987), the earliest record of a windmill is provided by a charter of St Mary's at Swinehead, Lincolnshire, dated 1170, and thereafter windmills spread so rapidly that in 1195 Pope Celestine III decreed that they should pay tithes. The Persians were also to develop a windmill, but it was the European windmill that spread east, not the other way round. In his *L'Estoire de la guerre sainte* of c.1190 Ambroise d'Evreux wrote:

> The German soldiers used their skill
> To build the very first windmill
> That Syria had ever known.

Tacking at sea was another Dark Age advance. The Roman square-rigged ships could not move upwind efficiently, so their sailors had to row, but the fore-and-aft or 'lateen' sail can tack, and its first representations appear in Greek manuscripts of the ninth century (the word lateen is cultural theft: the Latins did not invent the lateen sail, which was probably a Saracen invention). And Dark Age Europe continued to make inventions including cloisonné jewellery, felt, trousers, skis, butter, rye, hops and barrels. The Western world's first lock gates (which were invented independently of Chiao Wei-Yo's in China) were installed in Bruges, Belgium, in 1180.

Where did this extraordinary Dark Age technological inventiveness come from? There were no government-funded research labs then, so Francis Bacon's linear model cannot account for the innovations. But there was trade. Coins provide hard evidence of that and, as Philip Grierson showed in *Dark Age Numismatics* (1979), all of Europe (with the embarrassing exception of the islands of Great Britain) continued to mint coins after the departure of the Romans. Throughout the so-called barbarian societies of mainland Europe, therefore, the volume of trade may have decreased but it did not disappear.

And trade soon regrew. By the late 700s a seventy-year truce between the Christians and Arabs had revived the old imperial trade between southern Italy, Sicily and North Africa in timber, furniture, dried fruits, linen, wine and cheese. At around the same time the Viking trade of Eastern gold for European slaves, via the Baltic and Volga, was accelerating.

Where there is trade there will be entrepreneurs, and where entrepreneurs compete they will invest effort in the invention of better products, the better to undercut the competition, which during the so-called Dark Ages they did, generating the technological revolution of that era. It was that superior technology that provided the economic surplus off which, ultimately, Plutarch's precious Renaissance was to feed, but instead of thanking the market for liberating humanity from the degradations of Rome, Plutarch opted instead for abuse. This is suspicious, suggesting that Plutarch, as an intellectual, was more concerned with his governmental perquisites than with the needs of society at large.

THE MYTH OF THE AGRICULTURAL REVOLUTION

The Dark Ages also witnessed an agricultural revolution, so let me explode another myth, that of the 'classical' Agricultural Revolution of the seventeenth and eighteenth centuries. We are raised on heroic stories of pioneering Dutch and British farmers of the seventeenth and eighteenth centuries transforming the production of food, but these stories are misleading.

The conventional story of the Agricultural Revolution

This is easily told: Dark Age agricultural productivity was horribly low, wheat yields during the eighth and ninth centuries being, for example, only around 300 to 400 kilograms per hectare (approximately 300 pounds per acre), barely enough to sustain the workforce in producing them. During the eighth and ninth centuries, however, advances in horse technology powered heavy ploughing and three-field rotation, thus allowing wheat yields to double to around 800 kilograms per hectare.

For a thousand or so years thereafter, nothing much changed until, during the seventeenth and eighteenth centuries, technical advances in Holland and Britain – of which the most important was the introduction of clover and ryegrass to enrich the soil by 'fixing' nitrogen – allowed wheat yields to leap to around 2 tonnes per hectare, that being the average English wheat yield in 1910.

But because the conventional story conflates average with maximum yields it is factually correct but incomplete. Average wheat yields in Europe during the first millennium were indeed horribly low, some did indeed rise around the turn of the millennium, and they did indeed rise again during the canonical seventeenth- and eighteenth-century Agricultural Revolution. Moreover, the leaps in production did indeed coincide with the new technologies. But average yields are not the same as maximum yields, and when we examine those, a dramatically different story emerges.

Wheat yields on the royal farms in France during the eighth and ninth centuries were already 800 kilograms per hectare, double the medieval average, and by the late 1200s large farms in northern France and southern England were already yielding over 2 tonnes per hectare, the same as England's average yield in 1910! Around 1300, farms in the vicinity of St Omer and Arras were yielding 3 tonnes per hectare, 50 per cent greater than England's average in 1910!

How were those high yields obtained? They were rooted in the first and most important agricultural revolution, that of the New Stone Age.

The first (New Stone Age) agricultural revolution
It was during the New Stone and Iron Ages that ploughing, harrowing, hoeing, crop rotation and the fertilization of the soil with animal manure were invented, and it was during the New Stone Age that nitrogen-fixing legumes and pulses (peas, beans and lentils) were exploited after having been domesticated in the Fertile Crescent at the same time as cereals during the eighth millennium BC. (The Chinese, too, cultivated their own legume, the soy bean, which nearly 5,000 years ago one of the legendary Three Sovereigns, Shennong, celebrated as one of the five life-sustaining grains.)

The New Stone Age, moreover, invented two-field crop rotation, by which fields were planted with wheat, barley or rye only every other year. On alternate years the fields lay fallow, which allowed them to recover their fertility and which also provided livestock with food in the form of stubble.

The second (Dark Age) agricultural revolution
This emerged around the eighth century AD in the area roughly coinciding with the Flanders/Greater Paris region. There the three-field system of crop rotation was invented, by which only a third of the fields lay fallow at any one time. In the autumn one-third of the fields were planted with wheat, barley or rye, and in spring another third were planted with oats, barley and the nitrogen-fixing legumes.

The three-field rotation of the eighth century AD doubled agricultural productivity. Cleverly, it exploited the poor climate of northwest Europe, because spring planting works only when summers are wet, and, by providing two harvests a year, it reduced the risks of any particular crop failure leading to famine. Further, it provided oats to feed horses, thus facilitating the substitution of horse power for ox power.

The third (canonical) agricultural revolution

During the seventeenth century, during the 'famous' Agricultural Revolution, the Norfolk four-course system was invented in East Anglia in England. That represented two innovations. First, nitrogen-fixing clover replaced the nitrogen-fixing legumes; and, second, turnips, which were already an old-established crop, were widely introduced to feed livestock during the winter. Thus the fallow year was converted into proper pasture, to produce a four-course of wheat planted the first year, turnips the second, and barley the third. And because the barley was undersown with clover and ryegrass, those were grazed by cattle during the fourth year.

Yet the four-course system of the seventeenth and eighteenth centuries represented no miracle: it improved agricultural productivity by over 30 per cent, but it did not represent the introduction of nitrogen-fixing into farming; it coincided only with the replacement of the nitrogen-fixing pulses by the nitrogen-fixing clovers and with the wider spread of turnips. Which brings us to . . .

The great mystery of medieval agriculture

If the Dark Age agricultural revolution could provide the astonishing maximum yields of the London/Paris/Flanders regions, why were its average yields so low? And low they were: academic papers are rarely exciting but it is impossible to read W.H. Long's article 'The Low

Yields of Corn in Medieval England' without astonishment.[2] The average medieval farmer did not systematically weed his fields, nor did he harrow or hoe them, so the weeds starved his crop by competing for nourishment and sunlight. Even worse, the average farmer did not bother to undersow with pulses, nor did he fertilize with manure. Those neglects helped to keep down the weeds, but they also deprived his crops.

Yet the agriculturalists of northern France, Flanders and southern England were paragons, sowing and undersowing up to 40 per cent of arable land with beans, peas and vetches, weeding, harrowing and hoeing, and fertilizing with animal manure. Why were *they* so industrious? Because they served a market. The farmers close to the conurbations sold their produce to townspeople, so they stretched the technology to meet the demand.

The average peasant isolated in the depths of the countryside was a subsistence farmer. He served no market, because he couldn't reach one. Because he served no market, he did not specialize. He was a generalist, and his life was largely dedicated to the many tasks of self-sufficiency, of which food production was only one.

But the city-serving farmer specialized. He grew food, period. And as the market for food grew, so did his productivity. Between 1300 and 1800 the population of Paris rose from 100,000 to 600,000, yet because transport was so poor those people were fed by the produce of the same nearby farms. Those farms did not need much better technologies over that half-millennium – wheat strains, for example, were only some 20 per cent more productive in 1800 than in 1300[3] – because the local farmers could meet demand by applying existing technologies more intensively. In parts of Flanders some fields received up to eighteen cultivations during the course of a rotation, and in the mid-fourteenth century Norfolk farmers ploughed and harrowed up to six times.

Agricultural productivity during the medieval period, therefore, like technological growth during all periods of history, was driven by the market. This explains why, incidentally, England's canonical agricultural revolution was centred on East Anglia. Because transport by sea was then easier than by land, the farmers of Norfolk and Suffolk had long been important suppliers of the growing urban markets of the Netherlands and northern France.

DAVID RICARDO

Some economic historians understand that technology is the daughter of the market. So the current edition of the standard economists' dictionary, *The New Palgrave*, concedes under the entry on feudalism, that: 'Under the pressures of competition . . . [medieval] market-dependent farmers were obliged to adopt techniques which had long been available but long eschewed by peasants.' But many economic historians invert the story, seeing technology as the driver of markets rather than as the follower. Consider the English production of wool. Between the early 1300s and the mid-1400s sheep flocks in England shrivelled. Fleece weights fell by 34 per cent, and exports of English wool fell from around 40,000 to 10,000 sacks annually. The standard explanation for those disastrous events is that poor weather encouraged invasions of liver flukes, scab, murrain and other sheep diseases that crippled the animals. Nonsense!

If diseases had decimated the flocks, the price of the remaining wool would have risen because of unmet demand. But between the early 1300s and the mid-1400s its price fell by 50 per cent. Why? Because the Black Death and the Hundred Years War so disrupted ordinary life that they destroyed the demand for wool – that's why production fell. The fall in the price of wool, moreover, destroyed the farmers' incentive to maintain their flocks healthily or to sustain specialized breeds. But had the market remained good, the weather and the flukes would have been overcome.

Similarly, people are still taught that the Agricultural and Industrial Revolutions of the seventeenth, eighteenth and nineteenth centuries represented technologically driven increases in wealth. Actually, they represented market-driven advances in demand, which called for more technology, which then – and only then – produced the wealth.

It was David Ricardo (1772–1823) who propagated the myth that technology, not the market, limits economic growth. Ricardo, a London stockbroker with an interest in economics, understood that medieval Europe starved, and he therefore assumed that the demand for food must have been enormous. Since that demand was transparently not met, Ricardo concluded in one of his famous aphorisms that 'money cannot call forth goods'. He maintained that the farmers couldn't improve their productivity or their technology, or they surely would have: 'no improvements take place in agriculture' (*Essay on Profits*, 1815).

But Ricardo failed to understand that in medieval times transport barely existed. It was during the Middle Ages that the word travel emerged out of the Old French word *travailler*, which itself derived from the Latin word *tripaliare*, meaning to torture, because travel then really was torture. Roads then were truly terrible and often unsafe, and if people did not live within a day's walk of a marketplace then, effectively, they lived outside the market economy. Consequently, a rich family might have been starving in Paris, but its money couldn't call forth agricultural goods from the Loire valley if transport from the Loire was non-existent.

When, therefore, the Agricultural and Industrial Revolutions just 'happened', Ricardo could only explain them away as 'technological miracles'. To this day, too many economists see those two revolutions as scientific and technological triumphs rather than as the political, legal and economic achievements of the market, and too few commentators understand that science and technology follow the market; they do not create it.

Consequently, agricultural productivity started to accelerate in the greater Paris/Flanders region during the eighth century because politics and statescraft fostered a market. It was there that Charlemagne and other magnates established the law and infrastructure a market needs: those great men enforced contracts, they secured property rights, they maintained roads, and they executed footpads. They therefore provided the motivation for farmers to invest in new technology, which was why it was the farmers around Paris who invented horseshoes, harnesses, multiple-drawing, three-field crop rotation and the other technological advances of the day.

The Church, too, fostered the market. The Church then owned about a third of the fertile land of northern France and Flanders, and to protect its business the Church enforced its property rights and commercial law: it could afford lawyers, it ran its own courts, and it could afford to enforce the courts' judgements. We see, therefore, that the recovery of the market in northern Europe during the so-called Dark Ages resurrected innovation in agriculture, thus confirming that it is the market that calls forth new goods.

* * *

The so-called Dark Ages were not dark at all; they fostered the market by which new and wonderful technologies emerged.

7
Making Markets

'In democracy it's your vote that counts; in feudalism it's your count that votes.' Old Joke

Where markets survived in medieval Europe, people could be amusingly commercial in their thinking, concerned to avoid free-riding. The Cistercian monks were the great technologists of the day, and their abbey at Clairvaux was centred on the iron foundry. But the abbey also owned a watermill and this was how, during the twelfth century, the monks described it:

> The stream first throws itself at the mill, crushing the wheat under the millstones and shaking the sieve that separates the flour from the bran. At the next building the stream replenishes the vats and surrenders to the flames that will heat it to provide the monks with beer. Then, to provide the monks with clothes, its waters will lift and drop the heavy pestles of the fullers' hammers. At the tannery it will help prepare the leather. Finally, *in case it receives any reward for work not done*, with its dying efforts it will wash away all the waste, leaving the monastery spick and span [my italics].
>
> Quoted in Jean Gimpel, *The Medieval Machine*, 1992

But market thinking did not always flourish in medieval Europe because its necessary conditions – property rights, the rule of law and freedom to trade – were so frequently undermined by the chaos of the Dark Ages. Within certain parts of Europe – under Charlemagne and other magnates – stability did for a time emerge; but it did not last, central authority was lost, and much of medieval Europe sank into an anti-commercial ideology as theologically bizarre, as oppressive and as stagnant as any of the Bronze Age. We could call that ideology 'feudalism' or vassalage. There was nothing primitive about feudalism; it could have emerged only out of the detritus of an old and failed society.

FEUDALISM

Feudalism was a social contract, not a very nice one. At its heart was a mafiosi-style protection racket, with lords providing protection in exchange for fees. Indeed, feudalism got its name from *feu*, the Old French term for fee, reflecting the fees or *feus* the serfs paid their lords. But feudalism, like all self-respecting tyrannies, was anti-commercial, and because the village was an almost cash-free zone the *feus* were paid in time, not money. The lord allocated the serfs their land and, in exchange, the serfs worked the lord's fields for free, donating perhaps two days' labour a week. This was not an economic contract, it was – to stretch the word 'social' – a social one.

An early feudalism emerged as the Roman Empire declined, when its peoples coalesced on the great estates or *villas* (the word village derives from the *villas*, the country houses, of the aristocrats). Each estate was effectively autonomous, growing its own food and supporting its own craftsmen. Such estates provided rational responses to the chaos of the times. But classical feudalism in Europe accelerated around AD 1000, after Charlemagne had failed to sustain centralized states.

Because feudalism so oppressed the peasants, it had to elaborate a fierce theology that, by integrating the best of God and Plato, justified the oppression: the serfs were to be exploited by divine command (the word religion shares the same root as ligate and ligature, from the Latin *ligare*, to bind). During the fourteenth century, Master Thomas Wimbledon preached in a sermon at St Paul's Cross that:

> These three offices are necessary: priesthood, knighthood and labourers. To the priesthood it falls to cut away the dead branches of sin with the sword of their tongues. To the knighthood it falls to prevent wrongs and thefts being done . . . And to the labourers it falls to work with their bodies and by their sore sweat to get out of the earth the bodily sustenance for themselves and the others.[1]

Within its ideologies, medieval Europe was self-consciously moral. Arable land was distributed among villagers with scrupulous fairness, and each peasant worked a number of separate strips, distributed among different fields, so that good and bad soils were equally shared. Free trade was abhorred. Medieval Europe followed Thomas Aquinas in believing that the state should impose 'just prices' and 'just wages'. On 20 December 1315, for example, the king summoned a Parliament at Westminster:

> The archbishops, bishops, earls, barons etc presented a petition to the king and his council, praying, that a proclamation might be issued out settling the Price of Provisions in the manner following: 'Because, say they, that oxen, cows, muttons, hogs, geese, hens, capons, chickens, pidgeons, and eggs, were excessive dear, that the best ox, not fed with corn, should be sold for 16s and no more: and if he was fed with corn, then for 24s at most. The best live fat cow for 12s. A fat hog, of 2 years old, for 3s 4d. A fat weather or mutton, unshorn, for 20d, and shorn for 14d. A fat goose for 2½d. A fat capon for 2d. A fat hen for 1d. Two chickens for 1d. Four pidgeons for 1d and 24 eggs for 1d. And those

who would not sell the things for these rates were to forfeit them to the king.' Proclamation was made in every county of England accordingly.

English Parliamentary History for 1315[2]

Medieval kings distrusted the market but, in promulgating prices, kings were (unusually) sometimes being helpful – the medieval population being so small, there was sometimes only one seller or purchaser present at a transaction, so people needed to be told the market price.[3] Nonetheless, the underlying medieval spirit was anti-market. So, for example, intellectuals like Geoffrey Chaucer criticized lawyers for defending clients for money rather than pro bono. (In recognition of the medieval ideal, English barristers to this day cannot sue for their fees – which technically are only honoraria – and their gowns possess a little pocket into which clients are meant to slip their tips.) Doctors, too, were attacked for charging fees. Johannes de Mirfield of St Bartholomew's Hospital wrote:

> But the physician, if he should happen to be a good Christian (which rarely chances, for by their works they show themselves the disciples not of Christ but of Avicenna and of Galen) ought to cure a Christian patient without making even the slightest charge if the man is poor.
>
> *Breviarum Bartholomei, c.*1400

Where trade was tolerated it was tightly regulated because free markets were viewed as, literally, the work of the devil. How could it be right that, of two men working equally hard, one should earn more than the other simply because of vagaries of surplus or shortage? Pope Clement V's bull of 1312, therefore, confirmed that usurers would be consigned to eternal damnation. Medieval society recognized that loans sometimes needed to be made, but people believed that the rich should lend money for charity, not for interest: no man should enrich himself without labour simply by lending money at interest.

Because their contracts were social, not economic, feudal societies were rigid. The conditions of service in the fields were determined by ancient usage, and they obtained over generations. Feudalism, in short, presaged the huge, stable but sterile empires of the past. Like them, it enmeshed all humanity in a complete theocratic, cash-free, social tyranny. And, like them, it promoted no new technology, providing it with neither opportunity nor reward.

The outlook for science and technology in medieval Europe looked bleak, therefore, but for one saving: feudal Europe differed from past empires in one key respect; it was divided against itself. Kingdom against kingdom, province against province, village against village, the continent was perpetually aflame. Why?

It was the Marxist economist Maurice Dobb who showed why feudalism bred war: no matter how unfair, feudalism was at its heart a contract, which had emerged by the voluntary adherence of people to the local strong man.[4] Serfs and lords, therefore, were bound together by a network of mutual obligations under God, and serfs, as Christians, were not slaves, they enjoyed certain rights, no matter how restricted. Of those rights, the greatest was tenure. It was tenure that led to war.

Serfs were not free to leave the village: they were tied *glebe adscripti* to the soil. But on the other hand, the serfs couldn't be fired. No lord could remove them. Feudalism, therefore, deprived the lords of the one power they needed to persuade the serfs to work – the sack. Capitalism may be flawed, but when it replaced feudal relationships with economic contracts it also unleashed a powerful tool for motivating staff – the fear of dismissal. But feudal staff didn't fear downsizing, they were secure in their land tenure, and they didn't work at all hard. Indeed, medieval Europe recognized up to 150 holy days or holidays during which no work was done, and peasants often

SEX, SCIENCE AND PROFITS

spent the whole winter in bed. When they did put time into the lord's fields, they did as little work as possible.

And the lords hated the peasants for their indolence. The language the lords used about them reveals the depth of their abhorrence, the word villain being only a respelling of villein, a villager who lives on the lord's *villa*. Serf is derived from *servus*, the Latin for slave, which the serfs were not, but the name stuck because it was so insulting. And the savagery by which lords could treat the peasants speaks even louder of that loathing – witness the terrible scenes of the Hundred Years War.

Because a type of feudalism survived in Russia until relatively recently, the frustration it caused its landlords has been described in modern terms. In Tolstoy's *Anna Karenina* the character Levin found it impossible to motivate his peasants into adopting new techniques. Hating their feudal lord, the peasants simply refused to embrace Levin's well-meaning innovations. He soon grew to be less well-meaning.

In the Middle Ages, therefore, a greedy or ambitious lord had no commercial route to greater wealth. If he wanted to better his lot, he had no choice but to invade his neighbour and steal his wealth. In the absence of markets, cupidity found its outlet in predation. Thus did feudalism promote endless warfare. Yet a medieval Europe of innumerable competing statelets was almost Darwinian in its survival pressures, and medieval Europe selected states for their military prowess. Consequently, as another Marxist economist, Paul Sweezy, has shown, the continual warfare also selected, unpredictably perhaps, for trade.[5] War fostered, of all unexpected institutions, the market.

A lord, fighting for advantage, would need chainmail; but few medieval villages boasted of deposits of iron or of the expertise to turn iron ore into armour or weapons. So medieval lords had to trade; they had to sell food or wool for breastplates or swords. Medieval lords, therefore, had to foster a merchant class. Medieval kings, too, initially attempted to fight wars in the proper feudal spirit under which their subjects served in their armies from duty, but even medieval kings needed to purchase services that feudalism could not provide. Those might include a fleet or siege weapons or mercenaries. So medieval kings, too, had to foster a merchant class.

A merchant class, possessed of independent wealth, threatened royal prerogatives, but nonetheless kings grew dependent on the merchants, not only for their supply of *matériel* but also because merchants, even those engaged in non-military trade, grew rich and so provided a ready source of tax or loans. So those kings who refused to accommodate the merchants found their countries impoverished and their war chests depleted. They tended, therefore, to lose wars. Thus did markets spread by a form of Darwinism, red in tooth and claw.

And markets destroyed feudalism. Since Marxists delight in an era that was so class-based, many of the most thorough historians of feudalism have been marxists, and though their books may have had dreary titles (*Studies in the Development of Capitalism*, *Theories of Capitalist Development* and the like) their stories are compelling. It was the lords who broke feudalism: the peasants had tried to subvert it via repeated rebellions and abscondings from the villages, but the lords punished those. Yet as warfare promoted markets, so the lords themselves acted increasingly as profit-maximizers, ejecting or compulsorily buying off their serfs and replacing them with tenants or sheep and other sources of cash. And, as money increasingly circulated, so the lords' estates evolved into demesnes, where farmers owned or rented the land they farmed. Tenancies and demesnes were early features of agriculture in the Paris/Flanders regions.

Regardless of how cruelly they were created, the tenancies and demesnes – by incentivizing individuals to work in their own interests – were more productive than the lordly lands of listless, clock-watching feudal villagers. They thus fed the cities, which in their turn further promoted trade by purchasing 'liberties' or 'freedoms' from the king. Those might include the right to hold markets or to self-government, the latter being especially important because a government of merchants will be more trade friendly than one of lords. Thus did capitalism grow in a virtuous circle, to replace feudalism with a money-based society.

Thus, as is usual in social change, the transformation was led by the leaders of society. Children of the 1960s such as myself, who believed that social change comes from the bottom, are generally disappointed.

THE COMMERCIAL REVOLUTION

The first European country to break away from feudalism was Italy. Italy was well placed. Straddling the trading crossroad that was the Mediterranean, its international commerce was boosted as early as the late 700s by a truce between the Christians and Arabs. Italy was also mountainous, which helped protect the nascent city-states from their feudal overlords. Consequently, although they generally needed to fight at least one battle of liberation, the city-states, enriched by their 'liberties', found it relatively easy to fight off their feudal overlords, transforming themselves into independent mercantile republics.

As republics, they could settle to the business of commerce. First, though, they had to invent the tools of international trade, and the invention of those financial instruments demanded as much originality and rigour as any invention in technology or science. Coins and bullion, for example, rarely provide an optimal currency. The early goods-carrying mule trains that crossed Italy would return with gold bullion or coins, and if they escaped the depredations of bandits they would cross other mule trains passing in the opposite direction, also burdened with gold. But the invention of cheques and of bills of exchange (effectively, post-dated cheques) and of deposit banking facilitated trade wonderfully: the first known foreign exchange contract was issued in Genoa in 1156 to allow two brothers who had borrowed 115 Genoese pounds to reimburse the bank's agents in Constantinople with 460 bezants within a month of their arrival.[6]

But the great, discreet, advantage of international bills of exchange was that they could be used also to finesse the prohibitions on usury. Just as medieval scientists and humanists had to disguise search as 'research' and original scholarship as 'renaissance', so medieval bankers and entrepreneurs hid their loans to each other as international bills of exchange. Because nobody knew the 'real' rate of exchange between different currencies, the bankers and entrepreneurs set the nominal exchange rate between two currencies to the true rate of interest of the loan. Thus did medieval businessmen cunningly introduce one of the key tools of a capitalist economy – the loaning of capital.

Eventually, as states increasingly borrowed money so their governments legitimized loans at interest to themselves, as Florence did in 1403. By 1545 a state such as England had also legalized private loans (at 10 per cent, though in his essay 'Op Usury' Francis Bacon argued for 5 per cent for individuals and 9 per cent for commerce). In the wake of these commercial developments, great banks like those of the Bardi, Peruzzi and Acciaiuolo arose to spread branches throughout Europe. The Florentine Medici bank in the time of Cosimo (1389–1464) had branches in Rome, Milan, Pisa, Venice, Avignon, Geneva, Bruges and London. And the first cheque in the world was written in Pisa during the fourteenth century.

Incidentally, the word bank comes from the Italian *banca*, meaning bench, because the early bankers sat on benches at the side of the markets; if a banker failed, his bench was ceremonially broken, so we

SEX, SCIENCE AND PROFITS

get bankrupt from *ruptus,* the Latin for break. The early Italian dominance of finance has left other traces on the language. The Bank of England, in London, is situated close to Lombard Street; and, until it was discontinued, the two shilling coin in Britain's pre-1970, pre-decimal currency was called the florin, a relict of the Florentine coin that once dominated European finance. Indeed, the symbol of the British pound, £, represents the l of *libra*, transmitted via the French *livre*.

Insurance protects trade, and a merchant is more likely to risk a ship if it can be insured against loss. The Greeks had developed a primitive form of marine insurance, known as a bottomry and respondentia bond, which consisted of a loan, repayable at a high premium if the ship returned with goods, but not repayable if the ship was lost. But by the twelfth century the Italians had invented premium insurance, the modern variety, under which a ship is insured for a fixed price. By 1400 two-thirds of Venetian maritime trade was underwritten in Florence.

And the Florentines invented double-entry bookkeeping, in itself only an aid to accounting. Under the double-entry system every transaction is entered into a ledger twice, first, as a credit or debit of money, and second, as a credit or debit of goods. The initial purpose of double-entry bookkeeping was to check the accounts – at the end of a period of time, the debits and credits should balance – but the long-term effect was to help invent the trading organization or firm as an entity separate from any individual. If an enterprise can accumulate debts or credits in its own right, even if only on paper, it assumes an entity of its own.

In addition, the Italians invented the patent. Italian merchants increasingly devoted themselves to developing new products, but they soon encountered an apparent disincentive: why should someone devote time or money to developing a better mousetrap if a competitor can then sell an identical copy? Why, indeed, should an inventor devote time or money to developing a better mousetrap if, not needing to recoup the costs of invention, the competitor can even undercut the inventor?

The argument (as I shall show later) is false: after all, did people fail to invent novel financial instruments, including cheques, bills of exchange, premium insurance and double-entry bookkeeping, for fear that others would, inevitably, use them? Of course not. Nonetheless, Italian inventors were soon to demand grants of monopoly, grants that were known as patents from *patere*, the Latin for lie open (as in patently obvious), because they were openly published as letters patent. The world's first patent was awarded by the Republic of Florence in 1421, and the world's first systematic patent laws were tabled in Venice in 1474. And the 1474 laws were modern in spirit, trading disclosure against private incentive:

> If provisions were made for the works and devices discovered by men
> of great genius, so that others who see them may not build them and
> take the inventor's honour away, more men would apply their genius
> . . . and build devices of great utility to our commonwealth.[7]

Later I shall explain why patents are a conspiracy against the public, but let me here acknowledge that medieval patents were important because they recognized, for the first time in history, the institutionalisation of commercially inspired research. We see with the invention of the patent that Italian markets had generated a sufficiently large mass of inventors or men of great genius to foster their conspiracy against the public.

And the men of great genius did indeed invent devices of great utility, though they were not always patentable. Spectacles were invented in Italy around 1285, and around 1230 a bright German had even invented the button! The Portuguese then spread the button eastwards, which is why the Japanese word for it, *botan*, is still the Portuguese one, *botão*.

REBIRTH OF SCIENCE

Commercial Italy also fostered the rebirth of science. Consider mathematics. The Arabs had adopted decimal numerals, and in 1202 Leonardo Fibonacci of Pisa (*c.*1170–*c.*1240) introduced them to Europe in his *Liber abaci*. Subsequently, it was the Europeans, not the Arabs, who progressed science. The ancients had solved linear and quadratic equations, but in the intervening centuries there had been no advance in solving cubic equations until, in 1510, Scipione del Ferro, a professor of mathematics at the University of Bologna, discovered the solution to the cubic equation of the form $x^3 + mx + n = 0$ where there is no simple x term.

In 1534 Nicciolò Tartaglia, a teacher of mathematics in Venice, found the solution to the cubic equation $x^3 + px^2 = n$ where there is no simple x term, and in 1545 Gerlamo Cardano, a lecturer in mathematics in Milan, published his *Ars magna*, the major mathematics book of the era, which incorporated the solution to the quartic equation discovered by his own, brilliant, servant, Lodovico Ferrari. (Cardano, though, was a man of his time and, as he described in his autobiographical *Book of my Life*, he was imprisoned by the Inquisition for casting the horoscope of Jesus Christ.).

How were these mathematical discoveries financed? By the market. Contrary to Bacon's fears, the market funded research. Three major institutions emerged, for profit, not for profit and philanthropic.

For-profit

Tartaglia was a freelance maths teacher, and there were many of those and many small private maths schools or *scuola d'abbaco* in Italy at the time because they were the MBA professors and business schools of their day. They taught the basic skills of the market to an increasingly commercial society, and young men, anxious to learn accountancy or navigation or the other relevant industrial skills, attended them. To advertise themselves and their *scuola d'abbaco*, therefore, the private mathematicians competed to publish original research, just as academics still do.

Not for profit

Del Ferro, meanwhile, taught at the University of Bologna, a conventional charity that enjoyed papal recognition. It survived, though, on students' fees (the Italian universities, of which Bologna was the first, having been founded in 1050, were actually created by the students, who employed the professors), so it competed in a market and its academics competed against each other to recruit students, which motivated them to publish. And being not for profit, the university erected no institutional barriers (other than those imposed by the Church) to unfettered publication. Thus we see how one of the great Western traditions, the publication of original research, was fostered by the market in education.

Philanthropic

Cardano, the greatest of the mathematicians, taught in Milan at the Piattine schools, which had been founded by a bequest of Tommaso Piatti. The Piattine bequest illustrates another aspect of the market, namely trust. Feudalism depended on oppression and superstition, but commerce depends on trust, so we find that market cultures foster the moral values that translate, socially, into donations and philanthropy. And Ferrari's later career illustrates another way by which market societies reward scholarship: patronage. Capitalism generated wealthy patrons, one of whom, Cardinal Ercole Gonzaga, regent of Matua, supported Ferrari's scholarship. Such patronage of science became institutionalized, and Galileo, for example, was supported by the Medicis.

* * *

In his 1965 book *The Rise of Christian Europe* Hugh Trevor-Roper attributed Portugal's achievements in the age of Henry the Navigator to its having largely escaped the curse of feudalism: being a thin, barren, maritime nation only recently liberated from Muslim rule, it had retained its market and bellicose imperatives, both of which fostered research. Certainly in the rest of Europe science died under feudalism but was restored under capitalism.

The late-medieval market was good for science, which flourished without government subsidies. Interestingly, the three types of institutional support for science that emerged in medieval Italy were similar to those we see today in the OECD countries, especially in the US and UK.

8

Early Monopolistic Capitalism

'The mean rapacity, the monopolizing spirit of the merchants and manufacturers, who neither are, nor ought to be, the rulers of mankind.' Adam Smith, *Wealth of Nations*

Poor Italy. Having pioneered the greatest per capita wealth in Europe, she soon found herself invaded by the Spanish, French and Austrians. Italy had been warned – Machiavelli devoted the last chapter of *The Prince* (1513) to urging the Italians to unite – but for centuries Italy was to groan under the yoke of foreign occupation.

Yet here is an odd story: the commercial and technical leaderships of Europe did not settle on any of Italy's conquerors. Spain, France and Austria may each have been, *in toto*, richer and more powerful than any one of Italy's city-states, but they never learned to raise GDP per capita the way the Italian city-states had. They were plunderers, not wealth creators.

The wealth creators were the Dutch and English, who pioneered the canonical Agricultural and Industrial Revolutions of the seventeenth, eighteenth and nineteenth centuries. Why them? The clue is found by using fuller titles for the countries that made the history. The Commercial Revolution was pioneered by the Florentine Republic, Venetian Republic and Genoese Republic, the canonical Agricultural Revolution was pioneered by the Dutch Republic, and the Industrial Revolution was pioneered by the English 'republic' (which England has effectively been since the Glorious Revolution of 1688). In an otherwise monarchical Europe, the continent's technological leadership was successively assumed by the handful of republics. Why was this?

The kingdoms of *ancien régime* Europe failed to grow much, economically, but not because they were still feudal. They were not. By the sixteenth century most of western Europe had embraced markets, tenancies and demesne agriculture; but the Europeans hadn't embraced free markets. Instead, royal Europe was stuck in early monopolistic capitalism. The merchants of the era were obsessed by monopoly: they had purchased their commercial rights from the kings and lords as monopolies, so they defended them. Here is Jocelin of Brakelond's contemporary account in his *Chronicle Concerning the Acts of Samson, Abbot of St Edmund's Monastery* of a twelfth-century struggle over monopoly:

> Herbert the Dean set up a windmill on Habardun, and when the Abbot heard he was so angry he could no longer eat or speak. The next day, after Mass, he ordered his Sacrist to destroy it immediately. But the Dean protested that he had the right to build on his own land, that the wind was free, and that he wanted only to grind his own corn. But the Abbot replied that 'I thank you as I would thank you if you had cut off both my feet. By God's face, I shall starve until that building is destroyed. You know that not even the King can change anything within the liberties of this town without the assent of the Abbot. And you know perfectly well that people will soon flock to your mill and that, they being free men, I will not be able to punish them.' So, on the advice of his son Master Stephen, the Dean pulled down his own mill, and when the Sacrist arrived he found it already destroyed.
>
> Quoted in J. Gimpel, *The Medieval Machine: The Industrial Revolution of the Middle Ages*, Pimlico, 1992

Monopolies, of course, damage an economy because by limiting competition they force up prices and destroy the incentive to innovate.

And monopolies damage other monopolies. Consider the Spanish shepherds' monopoly. In 1273 King Alfonso X organized the various local shepherds guilds or *mestas* into a single guild, the Honourable Assembly of the Mesta of the Shepherds of Castile:

> The motive was merely one of the king's financial embarrassments; he realized that it was much easier to assess taxes on livestock than on men, and formed the *mestas* into an organization that would provide considerable sums to the monarchy. In exchange for these taxes the herders wrested a series of privileges from Alfonso X, the most important of which was the extension of supervision over all migratory flocks, including stray animals, in the whole kingdom of Castille. This supervisory function was extended, in time, even to 'permanent' sheep.
>
> Jaime Vicens Vives, *An Economic History of Spain*,
> Princeton University Press, 1969, p.25.

The consequences were devastating for the development of Spanish agriculture, because the *mestas* acquired rights over everybody else's land:

> The Council of the Mesta by the sixteenth century had become a privileged institution, with protected routes across the kingdom, with its own itinerant legal staff and armed guards accompanying the annual flocks, with authority to override conflicting interests, to prevent the enclosure of fields in their path, empowered to engage in collective bargaining with the most powerful landowners, exempt from the payment of the *alcabala* and from municipal sales taxes. It had judicial powers and economic prerogatives which placed it beyond the reach of other institutions.
>
> Maurice Schwarzman, 'Background Factors in Spanish Economic Decline' in *Explorations in Entrepreneurial History*, April 1951

SEX, SCIENCE AND PROFITS

Had the abbot of Bury St Edmund's or the *mestas* of Spain been exceptional, wealth in Europe would have grown satisfactorily, but unfortunately Bury St Edmund's and the *mestas* were only typical; there were then thousands upon thousands of privileged institutions in Europe whose judicial powers and economic prerogatives placed them beyond the reach of others. Thus did all of Europe stagnate into an interdigitating network of monopolies.

All? Well, not entirely . . . one small community of indomitable lowlanders held out for free trade – the Dutch. The Swiss historian Jakob Burckhardt once claimed that 'history is geography', and though no one can reduce a complex national history solely to geography, it does appear that the Dutch created modern capitalism because theirs are low lands (polders) that shelter behind dykes.

Large parts of the Netherlands lie below the level of the sea. They can survive only if they maintain their dykes. Since it is much easier to breach a dyke than to build one, the only society that could have evolved in the Netherlands was one sufficiently harmonious that people cooperated. Feudalism could not maintain the dykes because the miserable hierarchy that was feudalism generated resentful people who would be only too happy to see the dykes breached, even if they drowned with them. Feudalism celebrated proud lords clashing in cruel display, but Holland's national myth is of the peasant boy who inserted his finger in the dyke.

DUTCH EXCEPTIONALISM

The Dutch really were exceptional. As the economic historian and Nobel Laureate Douglass North chronicled in *Institutions, Institutional Change and Economic Performance* (Cambridge, 1990), the political story of medieval Europe is one of parliament after parliament surrendering its powers to the king. The French Estates General, for example, surrendered its powers to King Charles VII (reigned 1422–61), and the Cortes of Castile surrendered its to Ferdinand and Isabella (reigned 1479–1516). But the burgers of the Netherlands refused to surrender their powers, and the States General of the Low Countries rebelled, successfully. It was because the Dutch preserved their parliament that they created the legal framework for economic growth.

The Low Country burgers of the States General were no saints, and they were as keen as any English abbot or Castilian *mesta* on their own monopolies. But they were also keen that other peoples' monopolies should not infringe their own. So, in an extended process of mutual negotiation, the burgers of the States General negotiated with each other a doctrine of property rights that guaranteed individual property rights but also prevented people from illegitimately extending their rights to infringe others'. Consequently, the people of the Netherlands created, almost by accident, the conditions for a free market, and one that enabled the Netherlands to pioneer the first sustained period of rapid economic growth in history. We call that period the Agricultural Revolution because it was on agriculture, which then dominated the economy, that the technical and organizational improvements were focused. But that Agricultural Revolution was built on the market, the world's first stock exchange having opened in Antwerp in 1460.

ENGLISH EXCEPTIONALISM

Similar processes to the Netherlands' soon followed in England. England, too, was once a land of legalized monopolies, absolute monarchs and arbitrary appropriations. Consider Morton's 'fork'. When Henry VII's coffers were depleted he would send his

Chancellor, Cardinal John Morton, to demand money with menaces from the magnates. The magnates would be compelled to entertain Morton. If their entertainment was lavish, the cardinal would demand money on the grounds of their wealth; if their entertainment was modest, he would demand money on the grounds they must be hoarding their money. Either way, the lords and merchants had to surrender huge sums to the Treasury – hence the 'fork', a Tudor Catch-22. When private wealth could be so arbitrarily confiscated, who would have invested in the England of the day?

Successive Stuart kings were no better. In *Forced Loans and English Politics* (1987) Richard Cust describes how the Stuarts would imprison the rich for failing to 'lend' them money on demand. And the Stuarts used the patent laws to invent new monopolies to sell. England's first recorded patent was granted to Eton by Henry VI in 1449 for the design of its stained glass windows, but as W. Price explained in his 1906 *English Patents of Monopoly*, the Stuarts soon found more lucrative patents. The crown issued patents to whole industries, including soap, tobacco and starch, thus allowing a single man to purchase each trade in its entirety. And the Stuarts were blatant thieves: in 1640 the government simply stole £130,000 that private merchants had placed for safety in the Tower of London; numerous bankruptcies ensued. But in 1688, sick of the never-ending royal thievery, the magnates of England, in their Glorious Revolution, established parliamentary sovereignty.

England, being part of an island, was lucky in its geography. The parliaments on the continent had surrendered their tax-raising powers for reasons of defence. When the Estates General surrendered its powers to Charles VII, he was fighting the Hundred Years War against the English and Burgundians, and he needed untrammelled authority. When the Cortes surrendered its power to Ferdinand and Isabella, the monarchs were struggling to suppress a hundred rogue feudal lords.

The parliamentarians of both nations understood that the monarchs needed free reign.

But once the battles were won, the continental crowns retained their untrammelled powers. Europe's tragedy was that, land borders being so permeable, its different nations could invade each other so easily. Consequently, the nations that prospered were those that were run as military states: continental kings soon learned that they needed to maintain armies at all times, which also made it easy for them to oppress their subjects. Consequently, continental Europe developed an absolutist culture of dirigisme, one that the intellectuals of the day applauded, despising the Netherlands and England as unfinished states.

Intellectuals dependent on government patronage will too easily laud government, and the conceit that England had aborted its development into a proper state was one that lasted for centuries. As late as 1869 Matthew Arnold (the poet and government schools inspector) was bemoaning in his *Culture and Anarchy* that:

> We English are in danger of drifting into anarchy. We have not the notion, so familiar on the Continent and to antiquity, of the State – the nation in its collective and corporate character, and trusted with stringent powers for the general advantage, and controlling individual wills in the name of an interest wider than that of individuals.

Yet, thanks to the protection of the seas, the English were privileged, privileged *not* to need a State with stringent powers for controlling individuals. We English could be free! Not fearing invasion from Europe, we could afford in 1688 to imitate the actions of the Dutch and to abolish absolutist regimes. Indeed, to prevent any king from oppressing us, we refused to maintain a standing army. So we English created in our Glorious Revolution of 1688 a Parliament that was

SEX, SCIENCE AND PROFITS

sovereign (under a constitutional monarchy imported from the Netherlands).

The Glorious Revolution is often overlooked because it was bloodless and instituted boring things like property rights and the rule of law, yet it was the Glorious Revolution that spawned the legal and financial institutions that created Britain's Agricultural and Industrial Revolutions.

PARLIAMENTS ARE GOOD FOR TECHNOLOGICAL REVOLUTIONS

In their parliaments, as they traded their monopolies against each others', the great men of the Netherlands and England also traded in an independent judiciary. They needed to ensure that, in their disputes with each other, the umpire at the bench was neutral. The Stuarts had dismissed judges who had adjudicated against them, and the English bench soon learned to obey royal commands. Consequently, nobody's rights were secure. But the parliamentarians of the Netherlands and England surrendered their own powers over the justices, agreeing that 'be ye ever so high, the Law is above you', not because they wanted to surrender their own powers but because that was the only way to ensure that the other man surrendered his.

And parliamentary rule was the only way to lower taxes. Between 1470 and 1540, after Ferdinand and Isabella had got their greedy paws on them, tax revenues in Spain rose no less than twenty-two fold. Yet so profligate was the monarchy that, despite repeated confiscations of private property, the crown went bankrupt in 1557, 1575, 1596, 1607, 1627 and 1647. The defaults on the loans crippled Spain's financial sector.

The situation in France was no better. Under a series of dirigistes such as Jean-Baptiste Colbert (1619–83) France raised vast taxes, delegating their raising to farmers-general who stripped the land like locusts. Consider the salt tax. The salt was prepared in the marshes of Nantes, but the producers were not allowed to sell it in the marketplace and were obliged to sell to the farmers-general at a fixed price. The salt was then shipped to coastal depots where it was packed, sealed and registered. Barges then carried the salt up the rivers of France to inland depots where it was inspected and re-registered. There, by wagon, the salt was transported to a third set of warehouses to be re-inspected and re-registered. Finally, the salt was taken to a fourth set of warehouses, the retail outlets or *greniers à sel*, where the consumers were allowed eventually to buy it – at a ten-fold mark-up.

Where taxes are high, private investment in science, technology and business will necessarily be low – people won't have the money. The great advantage of the Dutch and English parliaments was that, because they were taxing themselves, they kept the burden down. England, indeed, famously embraced laissez-faire or the 'night watchman state' in which tax rates were dramatically low and the government limited itself to little more but the maintenance of the navy, I suspect we do not need 'police', property rights and the rule of the law. For long periods in England after 1688, for example, there was no income tax, and before 1900 Britain enjoyed overall tax rates of less than 10 per cent of national income.

Yet sometimes even democracies have to raise their taxes, and for long periods during the Agricultural and Industrial Revolutions tax rates in England and the Netherlands were high, largely to pay for wars. But people actually paid them. Indeed, one of the consequences of the Glorious Revolution of 1688 was that, paradoxically perhaps, government income from loans and taxes soon quadrupled, from £1.8 million in 1688 to £7.9 million in 1697. Because the loans were made voluntarily to governments that people trusted, and because the taxes were raised by a representative parliament, the raising of this money caused relatively little disturbance, and the money allowed the English to win their wars against the French and the Dutch.

Under democracy, taxes do not necessarily frustrate economic growth. This is partly because expenditure under parliamentary rule is held to account, but also because parliaments defend property rights: citizens know that, once they have paid their taxes, their residual money and their residual property are safely theirs. They can still invest. But subjects in France or Spain could not. One curse of continental taxes was that, under autocracy, their collection was frightful. In France in 1760, for example, there were no fewer than 30,000 tax farmers-general, 22,000 of whom were paramilitary police, uniformed, armed and possessed of almost unlimited powers of entry, search, detention and confiscation.

And they needed their powers: the salt tax alone was so resented that smuggling was a major activity. Jacques Necker, director-general of finance from 1776 to 1781, estimated that at any time up to 60,000 people were engaged in smuggling salt. Those smugglers needed to be caught, tried, imprisoned and executed, and between 1780 and 1783, for example, in just the one region of Angers, 2,342 men, 896 women and 201 children were convicted. And industrial monopolies were defended savagely: Eli Hecksher reported in his book *Mercantilism* (1934) how, in just one decade, Colbert executed no fewer than 16,000 unauthorized importers or manufacturers of cotton cloth. In the town of Vallence, for example, seventy-seven people were hanged, fifty-eight were broken on the wheel and 631 were sentenced to the galleys.

But in the Netherlands and England, under consensual rule, a wonderful thing happened – not only did government income actually rise to meet the needs of a modernizing state, but interest rates came down. Interest rates in England fell from 14 per cent in 1693 to 3 per cent in 1731. When capital becomes cheap, people will invest and economies will grow. In the words of Douglass North:

> The striking decline in interest rates in the Dutch capital market in the seventeenth century and in the English capital market in the early eighteenth century provides evidence of the increasing security of property rights.
>
> *Institutions, Institutional Change and Economic Performance*,
> Cambridge University Press, 1990

The rates came down because parliamentary England, unlike absolutist France, had created the legal framework by which financial institutions could flourish. As P.G.M. Dickson showed in his 1967 *Financial Revolution of England*, the constitutional and legal settlement of 1688 allowed men to establish, among other financial institutions, the Bank of England (founded in 1694), private banks (twenty-five by 1720, fifty by 1770 and seventy by 1800) and a stock market (£300,000 of securities traded in 1690, £15,000,000 in 1710).

So we see that the canonical Agricultural and Industrial Revolutions happened in the Netherlands and Britain because those two countries had created the political and legal framework by which people could invest. The beginnings of Britain's technological greatness can be precisely dated to the Glorious Revolution of 1688, not because after 1688 the government invested in research (it didn't) but because it fostered a free market.

* * *

Humanity's great battle of the last 10,000 or so years has been the battle against monopoly. Priests and emperors have perennially invented justifications for their unjustifiable perquisites, and every time humanity takes a step forward (the New Stone Age, early Greek civilization, the Dark Ages, the Commercial Revolution) the monopolists have fought back (the Bronze Age, the Roman, Chinese and other empires, feudalism, early monopolistic capitalism) because the advantages of monopoly are so great that entrepreneurs will always seek it.

Today the battle is being fought over patents, but that will be the subject of a later chapter.

Book 2
Recent History

9

The So-called Agricultural Revolution

'We have more machinery of government than is necessary, too many
parasites living on the labour of the industrious.'

Thomas Jefferson to William Ludlow, 1824

After 1688 we enter the modern world of representative parliaments,
the rule of law, property rights and markets. And a good thing too: life
before markets were free was horrible.

The first systematic survey of living standards in England, *Natural
and Political Observations and Conclusions upon the State and
Condition of England, 1696*, was written by Gregory King (1648–1712),
the pioneer statistician, though not actually published until 1801. As
secretary to the Commissioners for the Public Accounts, King had
access to official statistics, such as the hearth tax and excise returns,
and he also conducted his own surveys. He found that in 1688 there
were 5,450,000 people in England and Wales, but of those no fewer
than 2,825,000 starved (tabulated in **bold** in the Table 9.1).
Horrifyingly, over half of all English or Welsh people would have died
from malnutrition and poverty, King found, but for charity or poor
relief. Cruelly, even many people who earned wages – common
soldiers, common seamen, labourers and cottagers – needed
supplements.

Table 9.1 The Population and Wealth of England in 1688

Rank	Numbers of people	Average annual income per per person	Average annual expenditure per person
Aristocrats and gentry	153,520	£40 18s	£37 16s
Professionals and officers	308,000	£11	£10 3s
Freeholders and farmers	1,730,000	£9 16s	£9 6s
Tradesmen and craftsmen	484,000	£13 12s	£12 6s
Common soldiers and sailors	**220,000**	**£6 16s**	**£7 10s**
Labourers	**1,275,000**	**£4 5s 7d**	**£4 12s**
Cottagers	**1,300,000**	**£2**	**£2 5s**
Vagrants	**30,000**	**£2**	**£3**

These figures have been calculated from Gregory King's table as reproduced by P. Mathias (*The First Industrial Nation*, Methuen, London, 1969). The numbers of people increasing the wealth of the country were 2,675,520, but the numbers reducing it were **2,825,500** (in **bold**).

As the table shows, the fundamental problem was not one of distribution – had the aristocrats' income been equally distributed it would not have made much difference – but of poverty. There simply was not enough wealth to share among people. In years of good harvests, cheap food kept the mass of people alive, but in years of bad harvests the price of food drove people into the abyss. Yet, grim though life was in England at the end of the seventeenth century, it was better than it might have been. The average income per head in Africa today south of the Sahara is still lower than was England's in 1700.[1]

England was rich because it was approaching a market economy. By 1688 it boasted 50,000 families of merchants, tradesmen or shopkeepers; 60,000 artisans and craftsmen; and 55,000 professionals. Some 20 per cent of England's national product was attributable to imports and exports, then a remarkable figure. Only one other country in Europe was more commercial – the Netherlands – and that was the only country richer than England. (Merchants enrich a country because they compete with R&D to improve their technologies; so the merchant who invents a better mousetrap steals a march on the merchants who fail to.)

Further, England was also learning to regulate its population. In feudal times the peasants had overbred as policy: serfs produced as many children as possible because children were their pensions; feudal children were an investment because, in a classic tragedy-of-the-commons, all children were accommodated with equal shares of the commonly owned land, so parents tried to grab as much as possible via their offspring.[2] The consequence was grinding poverty. But under the market, when people owned their land, they bred with circumspection, and England developed concepts such as primogeniture or of limiting their offspring to the 'heir and the spare' so that property would be conveyed to the next generation intact. Thus do property rights allow market societies to avoid Malthus's trap.

ENCLOSURE

Only with the wider liberation of the market after 1688 did standards of living finally start to rise in the sustained fashion we call the Agricultural Revolution. The driving force was 'enclosure', which was code for individual property rights. Under feudalism, the peasants who had worked the strips of land had not owned them, so they had not been incentivized to invest in new technology, but under enclosure the owners of the land also farmed it, so they invested.

Acquiring old knowledge

Technical revolutions proceed on two bases: the acquisition of established knowledge, and the creation of new knowledge. The Agricultural Revolution saw both. So, for example, the Earl of Haddington (1680–1735) wrote: 'I got a farmer and his family from Dorsetshire in hopes that he would instruct my people in the right way of inclosing, and teach them to manage grass seeds.'[3] Haddington was typical of many aristocrats including Charles 'Turnip' Townshend or Townsend (1674–1738), who grew rich by exploiting existing knowledge (adopting turnips as a winter feed for cattle in Townshend's case).

British farmers banded together to create institutions of knowledge-dissemination. Farming societies such as the Highland and Agricultural Society of Scotland (1785) or the Royal Agricultural Society of England (1830) were established. The Gordon's Mill Farming Club of Aberdeenshire, for example, helped introduce the Norfolk four-course rotation to northeast Scotland, and agricultural shows were held regularly throughout the country, their competitions helping promote the development of better breeds. These societies were voluntary, supported by their members, not the state.

Creating new knowledge

As markets were freed, so eighteenth- and nineteenth-century British agriculture became a ferment of 'improvements', with new types of cattle, such as Herefords, being bred, new machines, such as Jethro Tull's seed drill, being invented, and new strains of fodder crops being introduced. Individual entrepreneurs flourished: in 1843 the land-owner John Bennet Lawes, assisted by the chemist Joseph Henry Gilbert, turned his farm at Rothamsted into a laboratory, where he discovered, among other good things, superphosphate, the marvellous fertilizer that is produced when bones are treated with sulphuric acid. Meanwhile, and ultimately more importantly, agricultural science was promoted by voluntary societies, including the Royal College of Veterinary Surgeons (1844) and the Royal Agricultural College at Cirencester (1845).[4]

The scientific method

Farmers understood the scientific method: in June 1793 a Basingstoke farmer, Thomas Fleet, advertised a sheep drench to cure liver rot. He offered to test it against any infected herd, suggesting that the herd be divided into two equally sick groups, only one of which would receive the treatment: 'All the Sheep shall be together during the Experiment, and live in the same manner, the *Drenching* only excepted.'[5]

The consequence of this ferment of knowledge-transfer and research was wonderful. During the sixteenth and seventeenth centuries British agricultural productivity had increased by 30 per cent, but the eighteenth century witnessed a doubling in the rate of increase to 61 per cent (half attributable to the four-course system, half to the new technologies), and by 1850 agricultural productivity in Britain was increasing by 0.5 per cent a year, an unprecedented rate historically, the 'Golden Age of Agriculture'.[6]

But where was the government? Surely all these improvements could not have occurred without government funding? Actually, they could. Eighteenth and nineteenth-century England was laissez-faire. There was no government support for agricultural R&D. For a short time the government did fund a Board of Agriculture, but all that did was to sponsor between 1803 and 1813 an annual course of lectures by

Humphry Davy on agricultural chemistry. So dwarfed was the board by the privately funded societies that in 1822 it was disbanded, to general indifference. Thus ended the government's sole contribution to one of the world's most important technological revolutions.

It was the landowners who funded the science. During the eighteenth and nineteenth centuries gentlemen, as 'hobby farmers', competed to sponsor agricultural R&D. King George III ran a model farm at Windsor, and he rejoiced in his popular moniker, 'Farmer George'. Equally, George delighted in conferring the title of 'Royal' on the agricultural colleges and societies (though the epithet was only honorific; unlike in France, it carried no money). And the aristocracy's interest in agriculture translated into literature: the most famous of pigs must be the Empress of Blandings, Lord Emsworth's prize animal in P.G. Wodehouse's *Blandings Castle*.

The problem of ownership

Why did landowners do research? Surely the bulk of the benefits of their advances would have accrued to others, so why did they bother? As Professor E.L. Jones of La Trobe University wrote in the standard textbook, *The Economic History of Britain since 1700*: 'The central puzzle is the emergence of a British taste for hobby farming and agricultural improvements.'[7] It is a puzzle for which Professor Jones could provide no answer.

Professor Jones's was a fair question: the difficulty of recouping one's personal investment in agricultural research was illustrated by certain improvers who wanted an immediate commercial return. Consider the advertisement in *The Country Magazine* for 15 December 1788 where an inventor offered to reveal 'an entirely new mode of Winter Feeding Sheep', which could increase profits by over 50 per cent, if 1,000 persons, each subscribing 20 guineas, would club together to buy the secret from him. We do not know if the inventor found his 1,000 subscribers, but we do know that British agricultural research and development benefited from vast private resources. Why did people invest their own money in research?

To answer this question, let me examine the economics of agricultural research in today's third world, since it is easier to collect detailed data on present-day activities than on those of England's past, yet the agricultural circumstances are similar.

AGRICULTURAL RESEARCH IN INDIA TODAY

The two most important crops in India today are wheat and rice, and because these are global crops they have benefited from the international research of the so-called 'green revolution'. But almost as important to India are sorghum and pearl millet, which, being drought hardy, will grow on non-irrigated dry land. They are known as 'poor people's crops' because they are generally sown by farmers who are barely removed from subsistence agriculture. And, being regional, these crops have not attracted much international research: their improvement has been indigenous.

In the words of Carl Pray, the agricultural economist, there is a 'widespread belief that small-farmer subsistence agriculture in developing countries cannot sustain a commercial private breeding industry for food crops'. Yet in their survey of Indian agriculture Pray and his colleagues found that the major seventeen private seed companies in India invest significantly in R&D.[8] They spend an average of 4 per cent of seed sales on R&D, and between them they employ thirty-one scientists with PhDs, forty-five with MScs and many more with BScs or technical qualifications. Indeed, the private sector spends as much on seed R&D as do the governments of India through their national and state seed corporations.

Pray confirmed that the benefits of research flow mainly to others: of the rewards that accrue from private seed research, the seed manufacturers capture only 6 per cent. The remaining 94 per cent is reaped by farmers and dealers in terms of higher yields. To Professor E.L. Jones that would represent a problem: why invest in R&D if 94 per cent of its benefits go elsewhere? But Pray entitled his paper 'Private Research and Public Benefit' because he showed that the seed manufacturers are more than happy, since their rate of return on their investment in R&D is no less than 17 per cent (a lot more than they would have got had they put their money in the bank). If a researcher makes profits of 17 per cent annually by capturing only 6 per cent of the value of his or her research, why should the researcher care that the remaining 94 per cent is shared by those who buy his seeds?

And, anyway, who exactly shares in that 94 per cent? Assuming that the original researcher sells his new seeds to 1,000 farmers, then each one of those farmers will capture the grand total of 0.094 per cent of the benefits of the research. Is a manufacturer really concerned if any one of his customers captures 0.094 per cent of the benefits of his research? Indeed, perhaps it should be more: why should a farmer risk transfering his custom from an old and tried seed to a new and untried one unless they were to capture at least 0.094 per cent of the benefits of the original research?

To extrapolate back to eighteenth- and nineteenth-century England, therefore, if the hobby farmers profited at an annual rate of return of around 17 per cent on their investments in agricultural R&D (and from the wealth that men like 'Turnip' Townshend made, we can assume they did), then their investments would have made perfect sense: that others benefited even more need not have inhibited them.

But men like 'Turnip' Townshend were not just researchers, they were themselves also consumers of others' research. Research is a two-way street: advances in science may be difficult to achieve, but they are almost as difficult to access. Benefiting from the research of others is hard. How easily can you, dear reader, understand papers in agricultural chemistry? As I discuss later, the benefits from doing research do, in fact, accrue to the researchers because they – and only they – can understand other people's research. During the eighteenth century, therefore, the hobby farmers benefited by their research, because only by doing it could they educate themselves in other people's advances. It was only because Townshend researched his turnips that he could exploit the advances of other turnip researchers and vice versa.

And there is a third reason why it was Britain and its hobby farmers that created the Agricultural Revolution, whereas France, with its state-subsidized research establishments (see later) failed. In his survey of Indian R&D Pray found that the hybrids produced by private research yield more than those produced by the state's scientists – that is, he found that private sector R&D is more effective than state-funded R&D. It is only by pricing its seeds artificially low that the Indian governments find buyers. But Pray showed that those cheap seeds are a false economy; farmers get more yield to the rupee with private seeds. Even worse, Pray also found that, by artificially shrinking the private sector's market, the public sector's inferior R&D had displaced or 'crowded out' the private sector's superior R&D.

We see, therefore, that even though sorghum and pearl millet are near-subsistence crops, the market will support private money for high-quality research, and even more of that high-quality research would have been available had the state not intruded into seed research and displaced the private sector. Similar processes during the eighteenth and nineteenth centuries would explain why laissez-faire Britain led the world through the Agricultural Revolution despite France's huge state-funded programme in agricultural research.

This argument often persuades Dutch and British people, who remember their agricultural successes under laissez-faire, but most people are raised in countries like France or Germany, whose governments during the nineteenth century, on recognizing their countries had been left behind, funded agricultural education R&D to raise (successfully) their national agricultural productivity. It is therefore difficult for most people to appreciate that, actually, their governments were only helping their countries copy the advances made in laissez-faire Holland and Britain, a help that would in any case have been superfluous had the market been freed.

Indeed, we continue to see improvements in agriculture even when governments strive not to support them but actually to prevent them. Ralph Weisheit is a professor of criminal justice at Illinois State University, and he showed in his 1992 book *Domestic Marijuana, A Neglected Industry* that marijuana is, by a wide margin, the US's largest cash crop, employing over 100,000 commercial growers.[9] And its R&D is not neglected: private individuals, such as Ed Rosenthal, research into enhancing its cannabinoid content, publishing their findings in books such as *The Marijuana Grower's Handbook*. Most of Rosenthal's books, however, have to ordered directly from the author at www.quicktrading.com because he is forever being prosecuted by the federal authorities.

Research into cultivating marijuana does not enjoy federal subsidies, yet ever stronger and ever more disease-resistant strains of marijuana continue to be developed, all by the (un)free market. Indeed, marijuana's cannabinoid content is now so high that it has become a seriously mind-altering drug.

* * *

As Bacon explained, ownership is key to myriad human incentives. Certainly the example of the Agricultural Revolution in Britain confirms one aspect of the importance of ownership: only after the farmers under enclosure owned the land they worked did they do the R&D that fuelled the canonical Agricultural Revolution. But since they did do the R&D, despite not retaining its full benefits to themselves, we must conclude that the ownership of ideas is different from the ownership of things, and that Bacon's concerns over market failure in R&D are unfounded, as we shall see later.

A NOTE ON THE STRUCTURE OF ECONOMIES

As we wrap up the Agricultural Revolution, one of its consequences is worth considering here: the growth of manufacture and then of services. Consider Gregory King's data in Table 10.1. In King's day 60 per cent of the population was engaged in agriculture, and only 15 per cent in manufacturing industry and 25 per cent in services. But, as Table 10.2 shows, by 1820 the percentage of the population working on the fields had fallen to 40 per cent. Today, that figure barely exceeds 1 per cent, yet that 1 per cent produces about 1,000 times as much food per worker per hour as did the 60 per cent in 1700, and it generates food mountains where the 60 per cent generated starvation.

Table 10.2 Structure of the UK Economy

Year	Agriculture %	Industry %	Services %
1700	60	15	25
1820	40	30	30
1890	16	44	40
1979	2	39	59
2000	1	25	74

Historic data comes from A. Maddison's *Phases of Capitalist Development* (Oxford University Press, 1982). Data for 2000 comes from *The Economist Pocket World in Figures 2003 Edition*, 2002.

When, during the eighteenth century, the percentage of people working on the fields started to fall and the percentage working in manufacture rose, earnest commentators predicted mass starvation: was not agriculture the only 'real' part of the economy? Similarly, during the twentieth century, when the percentage of people working in manufacture started to fall and the percentage working in services rose, earnest commentators predicted economic collapse: was not manufacture the only 'real' part of the economy? Today we understand that production, whether of food or widgets, is only a means to an end: what we want is services. We don't want our workforce stuck in mines, factories or fields, we want it on the high street cutting our hair and serving us coffee; we want it in schools teaching our children or in hospitals treating our sick or in libraries writing interesting books like this one.

As agriculture during the eighteenth century became more efficient thanks to the new technologies, and as the price of food consequently fell, so consumers could increasingly afford to buy manufactured objects. Consequently, the workers migrated into the factories, where salaries were higher. And as manufacture during the twentieth century got more efficient thanks to the new technologies, so the price mechanism impelled the workforce into the service industries: with artefacts becoming ever cheaper, customers could increasingly afford cappuccinos and other services.

Extrapolating into the future, we can predict the upcoming economic structure of lead countries. As agriculture becomes ever more efficient, fewer than 1 per cent of the population will work in that sector. Manufacture is going the same way, and as that becomes ever more efficient, so it will eventually employ only 2 per cent, say, of the population (a bright 2 per cent) who will programme computers that will control robots that will control other robots that will make things. Eventually, 97 per cent of the population will be in services, living off the supremely efficient manufacture and agriculture of the day. That 97 per cent will be dynamic, though, because even today 25 per cent of all R&D takes place in the service sector (a percentage that is rising), so the service sector itself is becoming ever more effective. R&D is itself a service industry.

As we shall see, lead economies grow at about 2 per cent per year, which means that they create some 2 per cent new jobs every year in new technologies (such as IT) and render some 2 per cent of old jobs (such as buggy whip manufacture) redundant. Economic growth, therefore, comes at the price of redundancy, and a civilized society helps the victims of growth to retrain – but that is separate matter.

10

The Industrial Revolution

'Virtually all economic activity in the contemporary world is carried
out not by individuals but by organizations that require a high degree
of social cooperation.' Francis Fukuyama, *Trust*, 1995

Go anywhere in the world today and should you encounter, of a
Friday or Saturday night, a youth vomiting in the gutter, he will
probably be a Brit. We Brits, in our drunken excesses, are the world's
louts. 'Twas ever thus, and Hogarth's print of eighteenth-century
Gin Lane in London merely depicted reality. Yet some commentators
have suggested that the seeds of British greatness may have lain in
our drunken loutishness, for louts are rebels, and drunks are trying
to forget. In *The Origins of English Individualism* (1978) Alan
Macfarlane, the historian and anthropologist, proposed that the
Agricultural and Industrial Revolutions had flourished in England
precisely because the English had always been rebels.[1]

Macfarlane expressed his thesis, as is proper for a historian and
anthropologist, in the language of property rights, showing in a survey
of historical records that, as early as 1400, individual peasants in
England were riding the demesene revolution and already carving out
those rights for themselves. When most continental Europeans were
simply units of an extended family, labouring on the family's holdings
and marrying those whom the family arranged, individual English
serfs were already owning property and were already choosing their
own spouses. They were thus developing the bloody-minded
individuality that characterized the English when their Peasants

164

Revolted, when they executed Charles II and when they Revolted again but Gloriously, Agriculturally and Industrially.

But the Industrial Revolution had its downside: it condemned the workers to wage slavery in factories. The craftsmen and craftswomen of Old England, no matter how poor, had worked at their own pace. Even as serfs, they had worked from home or in family-scaled workshops. But the Industrial Revolution spawned complex technology, which spawned the factories. There had long been some factories ('manufactories') in Europe, the most famous of which had been the Arsenal in Venice. Created in 1110, by the 1400s the Arsenal was employing over 2,000 men. During the crisis of 1571, as Venice struggled against the Turks, those men built no fewer than 100 ships in 100 days. But the Arsenal, being military, was an exceptional institution in a continent that was still largely agricultural: it required the Industrial Revolution to create the commercial factory as the instrument of mass employment.

Since the Industrial Revolution was British-born, so the first modern factory was British, John and Thomas Lombe's silk-throwing mill in Derby of 1719. By 1840 over half the manufacturing population of Britain was working in a factory. Yet the factories were no place for free-spirited individuals. In a factory, men and women are subordinated to the employer's machines, to work on his terms, at his times, under his direction, on his premises. Here is the list of fines imposed on errant workers in one cotton mill:

Any Spinner with his window open	1s 0d
Any Spinner washing himself	1s 0d
Any Spinner with an oil-can out of place	0s 0d
Any Spinner putting his gas out too soon	1s 0d
Any Spinner with his gaslight on too late in the morning	2s 0d
Any Spinner whistling	1s 0d

Any Spinner five minutes late	2s 0d
Any Spinner being sick and not finding a replacement	0s 6d
Any two Spinners together in the necessary [toilet]	1s 0d

William Cobbett, *Political Register*, 17 November 1824

Paradoxically therefore, the very culture of English individual enterprise that created the factories also produced men and women who resented working in them. Yet people wanted the money. The factories paid well and, contrary to myth, men and women flocked to them. As even Friedrich Engels conceded in his 1845 *Condition of the Working Class in England*, ordinary people resented the Factory Acts, which limited the hours of work. They wanted to choose their own hours of work.

Thus, torn between their appreciation of good money and their hatred of regimentation, the British responded in their characteristic fashion: they worked hard during the week, but they drank themselves insensible over the weekend, taking off the Monday and sometimes also the Tuesday to recover (St Monday's and St Tuesday's) and working ever harder over the rest of the week to pay for the following weekend's debauch.

Classically, the British Industrial Revolution is dated from 1780 to 1860. Although those dates must be arbitrary, those eighty years in Britain did witness some momentous events. First, the population nearly tripled. There were 7.5 million Britons in 1780, and 23 million in 1860. That alone is a tremendous fact. Until the British Industrial Revolution, any society whose population had greatly grown had also invited famine, Malthusian-style. When Ireland's population rose from 4.2 million in 1791 to 8.3 million in 1845, the consequence was the terrible famine of 1845, because its subsistence economy had grown dependent on only one crop, the potato. Nearly 2 million people either died or emigrated, and by 1851 the population of Ireland had fallen to 6.6 million, to fall even lower by 1861 to 5.8 million.

But the Irish enjoyed no property rights. Their landlords were absentee Englishmen who supplied their Irish tenants with such short leases that any tenant who made improvements was promptly 'rewarded' with a higher rent. So the only investment open to the Irish was their children, and they bred recklessly. So the Irish were subsistence, not commercial farmers, and when the potato crop failed, they were helpless. The British mainlanders, however, did enjoy property rights, they were commercial farmers, and, though their population tripled, it was only because their standards of living had risen so dramatically that they could afford more children. Between 1780 and 1860 real wages more than doubled from £11 per head to £28 per head.[2]

That increase in wealth was shared by all social classes. As Gladstone boasted in his Budget speech of 17th April 1863: 'The average condition of the British labourer, we have the happiness to know, has improved during the last twenty years in a degree which we know to be extraordinary, and which we may almost pronounce to be unexampled in the history of any country and of any age.' Where did the increased wealth come from?

SOME EARLY ECONOMICS

In his 1848 *Principles of Political Economy* the philosopher and economist John Stuart Mill defined the sources of national wealth as: 'Labour, Capital and Land. The increase of production, therefore, depends on the increase either of the elements themselves, or of their productiveness.' Thanks to Mill, therefore, the economists agree that a country can enrich itself if it increases its labour force, builds up its capital investment, brings more land under cultivation, or its population learns to exploit those factors more effectively to improve productivity. And the economic historians have confirmed that though all four factors grew between 1780 and 1860, the major source of Britain's increased wealth was a marked increase in productivity.[3] We know, moreover, how that increase in productivity was achieved: it was achieved by the introduction of new technologies and new institutions.

Let us consider the new technologies. These were not shared by all industries: a saddler in 1860 worked little differently from one in 1780. But a cotton worker's life was transformed, because the productivity pathfinder of the Industrial Revolution was King Cotton. When it had once required 50,000 hours to spin 100 pounds of cotton, by the early 1800s it required only 135 hours. Productivity in cotton continued to rise by no less than 2.6 per cent annually between 1780 and 1860, and a piece of cotton that sold for 80 shillings in 1780 cost only 5 shillings in 1860. A typical worker in 1860 produced more than ten times as much finished cotton per hour than he, or commonly she, had in 1780.[4]

How were those miracles of productivity achieved? At the beginning of the eighteenth century cotton was a manual trade, the two major processes being each performed by hand. The first process, spinning, which involved the twining of myriad short individual cotton strands into one endless thread, was performed on a spinning wheel, generally by an unmarried cottage girl (hence spinster). Her techniques had barely changed from those of Sappho of Lesbos's heroine in her poem of *c.*600 BC:

> Mother, I cannot mind my wheel
> My fingers ache, my lips are dry
> Oh! If you felt the pain I feel!
> But oh, whoever felt as I?

The second process, weaving, was similar to knitting, though it was performed on a simple loom. It, too, had barely developed over the centuries, but it was weaving that witnessed the first great technical advance when in 1733 John Kay (1704-79) created his 'flying shuttle', a mechanized loom whose vertical threads were held apart while the shuttle was shot, mechanically, to lay the horizontal weave. Spinning was then mechanized when James Hargreaves (c.1720-78) invented the spinning jenny during the 1760s, and when Richard Arkwright (1732-92) in 1769 linked the jenny to a watermill to remove the need for muscle power. In 1779 Samuel Crompton (1753-1827) linked the spinning jenny and the flying shuttle in a hybrid machine called a 'mule' (because mules are hybrids), which the Rev. Edmund Cartwright (1743-1823) redesigned as a power loom in 1785. From those high points flowed a vast number of smaller technical improvements.

Here is a question: were the increases in productivity primarily a consequence of the great technical advances such as the spinning jenny or of the myriad small technical advances that innumerable workers and manufacturers made to their machines alongside the big advances? Romantically, we attribute the increases in industrial revolutionary productivity to the great individual innovations such as the jenny, but when the economists do their sums they show that the vast number of small technical improvements overwhelmed the economic impact of the big innovations. As the economist Mark Blaug concluded: 'Innovations are rarely dramatic breakthroughs . . . but rather small improvements in a new process or product in which genuine novelty and imitation-with-a-difference shade imperceptibly into one another.'[5] Or, as Robert Stephenson told his biographer Samuel Smiles: 'The locomotive is not the invention of one man but of a nation of mechanical engineers.'

And who made the myriad small innovations? In 1851 the House of

SEX, SCIENCE AND PROFITS

Lords established a Select Committee on Letters Patent to study this question, and the evidence of the expert witnesses was unanimous: just as in Adam Smith's day, it was the workers on the factory floor who made the discoveries.[6] As one industrialist said: 'Generally, inventions come from the operatives.' Another industrialist, A.J. Mundella, a Nottingham hosiery manufacturer, explained who had been responsible for his firm's twenty-five patents:

> Every invention we have made and patented (and some have created almost a revolution in the trade) has been the invention of overlookers, or ordinary working men, or skilled working mechanics, in every instance.

This pattern of worker-generated innovation was typical. On 2 December 1874 the Inventors' Institute held a workers' exhibition, and as *The Times* reported the next day: 'The amount of inventive genius was astonishing.' Thus we see how Industrial Revolutionary Britain owed little to the university science of Francis Bacon's linear model, and everything to the inventiveness of the factory floor.

Systematic inventiveness, though, requires institutional incentives, and Isambard Kingdom Brunel (1806–59) told the 1851 Committee that the usual practice was for a firm to reward a worker with between £1 and £5 for the average useful innovation (when £1–£2 was a weekly wage). Great innovations could command more. By 1883 a typical engineering company such as William Denny, the Clyde shipbuilder, was reporting that since 1870 it had recognized fifty claims with rewards of up to £10 each.

Such an innovation as the institutionalization of invention shows how, to many contemporaries, the Industrial Revolution was predominantly a social revolution, characterized by the invention of a new art: management. Adam Smith had been fascinated by management, noting that the factories had generated a new profession of men who earned wages by their 'inspection and direction'. And when the economists do their sums, they show that the social innovations spawned and dwarfed the technical ones – as contemporaries understood. Consider Richard Arkwright. We now remember him for powering the spinning jenny, but to Andrew Ure, the Professor of Chemistry at Glasgow University, Arkwright was primarily a social innovator:

> To devise and administer a successful code of factory discipline suited to the necessities of factory diligence was the Herculean enterprise, the noble achievement of Arkwright.
>
> A. Ure, *The Philosophy of Manufacturers*, 1835

Ure concluded that 'mechanical invention was not of itself sufficient to found a successful manufacture' and that, indeed, the reverse was as true: once a factory was organized soundly, with an appropriate system of reward for inventiveness, it would of itself generate the necessary mechanical invention to further its development.

The Industrial Revolution, therefore, was based on the innovations of the factory floor (not on Bacon's linear model), and its new technologies flowed out of the social innovation of the factory (not out of academic science). Yet even if we place social innovation at the heart of the Industrial Revolution, that revolution did produce big technical advances, so how were the big technical advances made?

INTELLECTUAL ADVANCE

The Industrial Revolution is widely believed to have represented the linear application of science to technology. Some scholars have disputed that simple view, but the conventional story, as told in the standard school textbooks or in the *Encyclopaedia Britannica*, assumes it. Yet it only takes a moment's reflection to doubt the story. Consider the flying shuttle or the spinning jenny. They were admirable machines, but did they incorporate anything more intellectual than ingenuity or common sense? Where was the science?

Since the flying shuttle or the spinning jenny or the organization of the factory clearly did not embody much science, the myths about the Industrial Revolution have skirted around the cotton industry to attach themselves to the steam engine. Ah, now there was science in action! Those huge pistons, those jets of smoke and steam, that vast power – oh, the ingenuity of man in harnessing science! So let us study the steam engine and explode a few myths.

The first commercial steam engine, Thomas Newcomen's, was installed in 1712 at Dudley Castle, Worcester. It was huge, expensive and inefficient, but it clearly met a need because by 1781 about 400 had been built in Britain, most of them devoted to pumping water out of coal mines. There may also have been as many on the continent, one having built at Königsberg (now Kaliningrad) as early as 1721.

Newcomen's engine was simple. It consisted of cylinder in which a piston rode up and down. Steam was admitted to one side of the piston, but: 'The working of a Newcomen engine was a clumsy and apparently a very painful process, accompanied by an extraordinary amount of wheezing, sighing, creaking and bumping' (Samuel Smiles *The Lives of George and Robert Stephenson*, 1874). To understand the noises made by Newcomen's machine, note that it was not, unlike later steam engines, a pushing machine but a pulling one. The power did not come from the pressure of the steam, it came from creating a vacuum: the steam was passed into the cylinder and, once the cylinder was full of steam, Newcomen cooled it with a jet of water. The steam condensed, which sucked down the piston, which did the work.

Whence did Newcomen get his idea? The *Encyclopaedia Britannica* has no doubt: 'The researches of a number of scientists, especially those of Robert Boyle of England with atmospheric pressure, of Otto Von Gueriche of Germany with a vacuum, and of the French Huguenot Denis Papin with pressure vessels, helped to equip practical technologists with the theoretical basis of steam power. Distressingly little is known about the manner in which this knowledge was assimilated by pioneers such as Thomas Newcomen, but it is inconceivable that they could have been ignorant of it'.

Inconceivable or not, pioneers like Newcomen were ignorant of it. The historian D.S.L. Cardwell established in 1963 in his *Steampower in the Eighteenth Century* that Newcomen (1663-1729), who was barely literate, was a humble provincial blacksmith and ironmonger who, stuck out in rural Devon, had no apparent contact with science or scientists. Newcomen did, however, have much contact with the tin mines in the neighbouring county of Cornwall, and he knew that they were frequently flooded. There was, unquestionably, a market for an effective pump.

Being an ironmonger, Newcomen knew all about pumps. He spent his working hours repairing and making cylinders, pistons and valves – they were his stocks in trade. Newcomen's idea, though unquestionably brilliant, was also very simple: he reversed the process, using the entry and exits of steam through a valved pump to turn it into an engine. Newcomen's idea, though it took him ten years of exhaustive experimentation to develop into a working machine, was the stuff of intuitive genius, but it was no more than the intuition of a creative man familiar with pumps, the domestic steam kettle and cold streams. No theoretical science was involved, nor was it necessary.

Newcomen, moreover, had not been the first practical engineer to discover that cooled steam could be harnessed. Thomas Savery (*c.*1650-1715), an English military engineer, had in 1698 patented a sucking steam engine (it did not work very well). Even earlier Salomon de Caus (1576–1626), a French Huguenot garden engineer who designed fountains, had used the condensation of steam to create the very first steam pump. To suck up water from wells, de Caus would light a fire under a boiler that contained a small amount of water. Attached to the boiler was a pipe that fed into a well and, when the water in the boiler had boiled into steam, de Caus would extinguish the fire. As the boiler cooled, so it would suck water up from the well. De Caus's success thus disproves the suggestion that the exploitation of steam power by engineers depended on developments in the science of vacuums, because he wrote his famous book *Les Raisons des forces mouvantes avec diverses machines* in 1615, yet the scientific study of vacuums did not start until thirty-five years later when Otto von Guericke started his academic researches.

Von Guericke (1602–86) was an interesting man. Having been an engineer in the Swedish army during the Thirty Years War, it was only on retirement, when he returned to his birthplace, Magdeburg, as mayor, that he was free to indulge his true love, research. His big step was, in 1650, to convert a water pump into an air pump, thus creating humanity's first vacuums (it was a big step because Aristotle had denied that vacuums could exist, and Rabelais had claimed that Nature abhorred a vacuum). Von Guericke studied vacuums, and he showed they would not conduct sound, only light. He also showed that vacuums would not support animal life (you can guess how he showed that). But he confirmed the power of vacuums in 1654 when, in a famous demonstration before the emperor and his court, he showed that it took sixteen horses on each side to pull apart two hemispherical vessels that contained a vacuum, yet those hemispheres were only 14 inches (35 cm) in diameter.

Von Guericke was a good gas scientist, as were his contemporaries, who included Robert Boyle, Christiaan Huygens and Denis Papin, but they were following, not leading, the engineers. It was because de Caus had sucked up water with a vacuum that von Guericke studied vacuums. The irrelevance of science to the technology of the engines of the day was illustrated by their next improvement. That came in 1764 when James Watt invented the separate condenser.

To cool the steam in his cylinder, Newcomen had cooled the entire cylinder, including the wall. That was wasteful because much of the heat energy of the steam was lost in re-heating the wall with each cycle. But Watt built a pipe into the cylinder and, at the appropriate times, the steam was passed down the pipe into a separate cylinder or condenser. This separate condenser was kept continuously cold, so it was only the steam, as it travelled from one cylinder to another, that was cooled.

Where did Watt get his idea? The conventional view is that Watt was applying Black's discovery of latent heat.

JOSEPH BLACK

Joseph Black (1728–99) was a doctor and scientist who, like Watt, worked in Glasgow. The two men knew and respected each other. Black was an easy man to respect for he was an engaging person and a skilled researcher, whose discoveries ranged over many disciplines. His first major discoveries were not even in physics but in bio-chemistry: he discovered carbon dioxide while he was studying bladder stones. Bladder stones are now rare, but they were common in Black's day and they could be excruciatingly painful. It was while studying how to dissolve them that Black showed that, on being heated, they gave off a gas or 'fixed air', which was to be characterized as carbon dioxide.

He also discovered latent heat. Latent heat is the heat you have to apply to ice to melt it or to water to boil it. Such heat does not raise the temperature of the material, it is the energy required to change the state of the material from solid to liquid or from liquid to gas. It is called latent because it is hidden in the physical state rather than the temperature of the material.

When, therefore, Watt (1736–1819), a rude mechanical in Glasgow, discovered how to reduce heat loss during the steam engine's cycle, educated gentlemen supposed that he must have been inspired by the discovery of latent heat made by Black, an educated gentleman in Glasgow. Here is an extract from J.D. Bernal's influential book *Science in History* (1954): 'The first practical application of the discovery of latent heat was to be made by a young Glasgow instrument maker, James Watt . . . [which] was crucial for the development of the steam engine.'[7] Except that Watt himself explicitly denied the suggestion:

> Professor Robinson, in the dedication to me of Dr Black's 'Lectures upon Chemistry', goes to the length of supposing me to have professed to owe my improvements upon the steam engine to the instructions and information I had received from that gentleman, which certainly was a misapprehension; as, although I have always felt and acknowledged my obligations to him for the information I had received from his conversation, and particularly for the knowledge of the doctrine of latent heat, I never did, nor *could*, consider my improvements as originating in those communications.

> Quoted in D.S.L. Cardwell, *Turning Points in Western Technology: A Study of Technology, Science and History*, New York, 1972

As Watt explained, his inventions:

> . . . proceeded solely on the old established fact that steam was condensed by the contact of cold bodies, and the later known one that

water boiled *in vacuo* at heat below 100°(F), and consequently that a vacuum could not be obtained unless the cylinder and its contents were cooled at every stroke to below that heat.

Moreover, as the scientific historian Cardwell pointed out in a devastating attack on the conventional view:

> The greater the latent heat in steam the *less* the incentive to invent a separate condenser, for since the heat required to warm up the cylinder is independent of the properties of steam it follows that the greater the latent heat the higher the proportion of useful steam supplied per cycle.

Watt's advances, therefore, owed nothing to contemporary science; they proceeded on an 'old established fact'. Watt, moreover, had not been formally educated in science; he worked on the Glasgow University campus as a self-employed instrument maker. And when he conceived his idea of the separate condenser he was working as a contracted technician (not as an academic) mending a broken Newcomen engine.

Watt was, indeed, working at the university for reasons that Adam Smith found exquisite. Watt had trained as an apprentice in London, but when he returned home to Glasgow to set himself up as a self-employed instrument maker, he was blocked by the civic guilds, which blocked autonomous entrepreneurs from trading. But the university was independent of the guilds and it could licence its own tradesmen. It needed an instrument maker, so it allowed Watt to set up his business on campus as long as he did the university's work when asked (for which items of work he was paid apiece, as was proper for an independent tradesman). Thus Watt's success owed nothing to Bacon's linear model and everything to Smith's free markets.

Moreover, the next major advance in steam engine technology, the use of high-pressure steam to push the piston, was made by a man in Newcomen's mould, Richard Trevithick (1791–1833). Trevithick was self-taught. Born in Cornwall to a mining family, Trevithick received no education other than that provided at his village primary school, whose master described him as 'disobedient, slow and obstinate'. But Trevithick addressed a problem. The Cornish tin mines were a long way from the nearest coalfields, so their Watt steam engines were expensive to run. Could they be made more efficient?

Trevithick thought so, and he used high–pressure steam to push, not suck, the piston. His high pressure steam engines were twice as efficient as Watt's, and they were so light that they could be transported around the Cornish mines. Nicknamed 'puffer whims' from the steam they emitted, over thirty were built, and some were converted to hoist ore as well as water. In 1803 Trevithick built the world's first steam railway locomotive at the Pennydarren Ironworks, South Wales, which on 21 February 1804 hauled 10 tons of iron and seventy men along 10 miles of track.

These dates of 1803 and 1804 are, sadly, significant: why did it take over a quarter of a century for high-pressure steam to replace Watt's sucking engines? The answer is that Watt's patents were extended in 1775 by a private Act of Parliament until 1800. For twenty-five years, therefore, all further development of the steam engine was blocked by a very litigious James Watt. Only after 1800 could developments like Trevithick's be made, and even then men like Trevithick continued to be harried by Watt. Consequently Trevithick, who, though a great engineer, was not as commercially shrewd as Watt, died a pauper on 22 April 1833. He was buried in an unmarked grave in Dartford, Kent. This is what Trevithick wrote shortly before his death:

Mr James Watt said that I deserved hanging for bringing into use the high pressure engine; this so far has been my reward from the public, but should this be all, I shall be satisfied by the great secret pleasure and laudable pride that I feel in my breast from having been the instrument of bringing forward and maturing new principles and new arrangements, to construct machines of boundless value to my country.

Quoted by Anthony Burton, *Richard Trevithick*, Aurum 2002.

The very next major advance in steam power after Trevithick's was made by yet another an ill-educated, barely literate, barely numerate self-taught artisan, George Stephenson (1781–1848). Light though it was, Trevithick's locomotive was still too heavy for the cast-iron rails of the day, so it needed further development. But on 27 September 1825 a steam engine designed by George Stephenson drew 450 people from Darlington to Stockton (at the frightening speed of 15 miles an hour). Stephenson went on to build the Liverpool to Manchester line, for which he designed *Rocket*, an engine that could attain 36 miles an hour! So incredible was its speed that on its inaugural run on 15 September 1830 *Rocket* ran over and killed a politician, the Rt Hon William Huskisson PC MP, who misjudged its terrific velocity.

Yet Stephenson was unschooled. The son of a mechanic, he followed his father in operating a Newcomen engine to pump out a coalmine in Newcastle. He only learned to read (just) at the age of nineteen when he attended night school, and he never really acquired mathematics. So dense was his Geordie brogue that, when talking to educated men from London, they needed an interpreter. Yet it was the educated men – from all over Europe – who consulted him, not the other way round.

So we see that the development of the steam engine, the one artefact that more than any other embodies the Industrial Revolution, owed little or nothing to science; it emerged from pre-existing technology and it was created by uneducated, often isolated, but commercially incentivized men who applied practical common sense and intuition to address the mechanical problems that beset them.

Looking back at the Industrial Revolution, it is hard to see how science might have offered much to technology, because science itself was so rudimentary. Chemists who subscribed to the phlogiston theory (that fire is a substance) or to the caloric theory (that heat is a substance) or who tried to build perpetual motion machines were not to be of use to engineers. Indeed, during much of the eighteenth and nineteenth centuries, the reverse was true; the scientists scrambled to catch up with the engineers. It was on Joseph Black's discovery of 'fixed air' that Lavoisier could show that fire represented oxidation, not phlogiston. The technology came first and the science followed – Francis Bacon's linear model is wrong.

And the steam engine itself drove the final nails in the caloric theory: it was Sadi Carnot (1796–1832) who, frustrated with Watt's steam or 'fire' engine, wrote the modern laws of thermodynamics. Watt's engine broke the rules of contemporary physics, being more efficient than theory predicted, so Carnot changed the theory, revealing the genesis of his theorizing in the very title of his 1824 book *Réflexions sur la puissance matrice du feu* (Thoughts on the Motive Power of Fire).

The irrelevance of academic science to technology during the eighteenth and nineteenth centuries can best be illustrated by comparing Britain and France. The Industrial Revolution happened in Britain, when its universities were notoriously stuporous. England only had two universities, Oxford and Cambridge, and we know what they were like. Glasgow was certainly better, but not all the Scottish universities thrived. When Dr Johnson visited St Andrew's University in 1773 he found one of its three colleges being demolished because no one attended it. When he tried to visit the library of another, no one could find the key. As this passage from his 1775 *Journey to the Western Isles of Scotland* shows, Dr Johnson understood the relationship between universities and wealth: he knew that universities grew out of national wealth, and not the other way round:

> It is surely not without just reproach, that a nation of which the commerce is hourly extending and the wealth increasing . . . suffers its universities to moulder into dust.

How badly Britain compared with France . . .

THE FUNDING OF RESEARCH

During the seventeenth century the British and French governments launched similar policies for science. Each country's king founded a similar body, the Royal Society in London (1662) and the *Academie des Sciences* in Paris (1666). The preamble of the statutes of Royal Society stated: 'The business of the Royal Society is: to improve the knowledge of natural things, and all useful arts, Manufactures, Mechanic practices, Engynes and Inventions by experiment.' The *Academie de Sciences* had similar aims. But thereafter the two countries diverged. After 1688 the British adopted laissez faire, and organizations like the Royal Society were left to sink or swim. It nearly sunk.

France embraced dirigisme. Following the direction of Colbert, chief minister to Louis XIV between 1665 and 1683, the State drove every aspect of French society, including trade, industry, education and science, the better to promote economic growth. The government raised tariffs to deter imports, it raised taxes to pay for bureaucrats to lead the economy, and it raised yet more taxes to pay for the schools of science and technology that were to provide economic growth. Colbert created the *Ecole de Rome* to teach arithmetic, geometry and draughtmanship, the *Academie Royale de Peinture* to teach painting, the *Ecole Royale Graduite de Dessin* to teach design, and the *Academie Royale d'Architecture* to teach architecture.

Colbert also fostered the great workshops of the Savonerie and the Gobelins, the mint, the royal press and royal manufactory in the Louvre. The *Academie des Sciences* received generous state aid, and the world's first scientific journal, the *Journal des Savants*, was created by the state. The *Jardin du Roi* was reorganized in 1671 to study botany and pharmacy. Three chemistry research laboratories were created by state, one in the king's library, one in the Louvre and one at the Observatoire.

Successive administrations continued Colbert's policies. The school of civil engineering or *Ecole des Ponts et Chaussées* was created in 1716, the *Ecole du Corps Royal du Genie* in 1749 and the *Ecole des Mines* in 1778. The *Ecole Polytechnique* was founded in 1795. And those institutions so thrived as to produce scientists of the quality of Lavoisier, Berthollet, Leblanc, Carnot, Monge, Cugnot, Coulomb, Lamarck, Cuvier, Geoffroy Saint-Hilaire, Gay-Lussac, Arago, Ampère, Laplace and Chaptal. Nor were the technicians ignored. Trade schools or *Ecoles des Art et Metiers* were founded all over the country, and by the early nineteenth century, when it was still only a craft in England, France had established engineering as a profession, with schools, formal examinations and, after 1853, its own research laboratories in the *Conservatoire*.

Yet it was laissez-faire Britain, whose laboratories and formal scientific education were pathetic, that fostered the Industrial Revolution, while France, which possessed the finest labs and research schools in the world, lagged economically. Why?

The Academies and Ecoles of France were fiendishly expensive. The appalling costs that France bore as it raised its taxes to pay for them might not have mattered had the money been well spent, but unfortunately it was not. That is the trouble with dirigisime; a centrally planned economy can work only as well as the plans, and France subscribed to the wrong plans.

Economic growth is, fundamentally, a matter of applying new techniques. In a free market innumerable entrepreneurs each chance their different innovations. Some entrepreneurs will invest in pure science, others will invest in applied science, and others will focus on marketing or on innovations in management. The market will then select the innovations it values. During the nineteenth century entrepreneurs in Britain offered the market a range of innovations, and the market opted for a judicious investment in applied science and a generous investment in the new management techniques.

But a centrally planned economy can make only one offering, and the French state opted for a huge investment in pure science. All the intellectuals recommended it. Yet the model was wrong. The Industrial Revolution did not represent the application of science to technology, it represented the development of pre-existing technology by hands-on technologists working in a market, and the important factor common to Newcomen, Trevithick and Stephenson, apart from that each was essentially illiterate and innumerate, was that each worked on pumps or steam engines as an artisan. They were not academics. Only Watt was employed in a university, but even he was working in a commercial capacity in a free market island.

Newcomen, Trevithick, Stephenson, Watt and their business partners, moreover, possessed the necessary capital. Science requires investment. A novel product cannot be brought to the marketplace until it has been researched and developed, which is expensive. Potential French entrepreneurs were taxed to extinction but, as Voltaire wrote in his 1733 *Lettres Anglais on philosophiques*, in England:

No one is tyrannized over, and everyone is easy. The feet of the peasants are not bruised by wooden shoes; they eat white bread, are well clothed and are not afraid of increasing their stock of cattle nor of tiling their houses, from any apprehension that their taxes will be raised the year following.

We can see, therefore, why Britain – not France – pioneered the Industrial Revolution. Intellectuals on both sides of the channel championed Bacon and the state funding of pure science, but intellectuals had no power in free-enterprise Britain where only the market ruled, so the British people invested their money in management and technology. The French state listened to its intellectuals, it raised taxes, and it dissipated its money on science.

One curious consequence of Britain's early but appropriate neglect of science was that its less successful competitors accused it of stealing their own, just as we once criticized Japan. A Swiss calico printer complained in 1766 of the English: 'They cannot boast of many inventions, but only of having perfected the inventions of others; whence comes the proverb that for a thing to be perfect it must be invented in France and worked out in England.'[8]

CHARLES BABBAGE

As the Industrial Revolution developed, so the British government continued to refuse to fund science, a reluctance that was strengthened by its unfortunate experience over Charles Babbage. A great myth has attached to Babbage, namely that he was a heroic genius and pioneer, the Father of the Computer, whose brilliance was denied us by the short-sightedness of the government.

Babbage was born in 1792, a year after the great Michael Faraday, but few contemporaries were more different. Babbage could never finish a job. The moment things got difficult he would spin off into public controversy. His entry in my edition of the *Chambers Biographical Dictionary* concludes, 'in his later years he was chiefly known as a fierce enemy of organ grinders', and that sums him up: he was a man who poured intense energy into bombastic conflict against trifles, while eschewing the dedicated labour of true achievement.

Babbage's was a conventional academic career, culminating in 1823 in the Lucasian Professorship of Mathematics at Cambridge, but during his tenure, unfortunately for everybody's peace of mind, he conceived of a calculating machine or 'difference engine'. The subsequent importance of the electronic computer has cast a rosy glow on Babbage's machine, but in reality it was little more than a mechanized abacus. The German inventor William Schickard had devised one during the 1620s, and though Babbage did conceive of a machine with memory, a programmer and a mechanical calculator, his ideas were inchoate. Yet few men possessed Babbage's mastery of public controversy, and by pamphleteering, lobbying and the issuing of dire threats of national decline, he eventually squeezed no less than £17,000 out of the government when the average wage was £2 a week.

After spending his £17,000 Babbage still had not build the engine, but rather than finish the job he announced that he now wished to create a completely new machine, the 'analytical engine', which employed punched cards. The government, having spent its money for nothing, refused to fund this new machine, which prompted Babbage into doing the one original thing in his life: he published *The Decline of Science in Britain*, thus inaugurating the tradition by which prominent scientists denounce the government for neglecting science. Yet with true Babbage incompetence he published the *Decline* in 1830, the year before Britain was to make two of the most important contributions to science that any country anywhere has ever made, for it was in 1831 that Michael Faraday discovered electromagnetic induction and it was in 1831 that Charles Darwin boarded HMS *Beagle*.

Those who still believe that Babbage was a misunderstood genius should consider the subsequent history of the difference engine. Twenty years after Babbage abandoned it, two Swedish engineers, Georg and Edvard Scheutz, obtained a grant from their own government to build it.[9] Being capable men, they had succeeded by 1853 for a cost of only £566. They then tried to sell the machine but, despite much hard marketing, managed to sell only two engines, each only to a government. Quite simply, contemporary mathematical tables did a better job. Private investors had long anticipated the engine's failure, which is why its only sponsors had been governments, organizations that are notoriously careless with taxpayers' money.

But Babbage's wasted £17,000 were, curiously, a blessing, because they warned successive British administrations off science. France's tragedy was that its government-funded scientists did nice work, but without a free market the French could never discover what a bad economic investment it all was.

THE INDUSTRIAL REVOLUTION'S BAD PR

The Industrial Revolution has not enjoyed a good press. One early critic was William Cobbett, whose *Rural Rides* (1830) described pre-Industrial Revolution England as a rural arcadia of ambling peasants whose happy exertions had been limited to little more than the plucking of plump apples from ample orchards. But tragically, Cobbett maintained, the cruel and callous enclosures drove the peasants off the land into the urban factories, there to be worked into early graves by cruel and callous masters.

Cobbett was no economic historian. Gregory King had already shown that during the seventeenth century agricultural labourers were the poorest section of society, and by the nineteenth century agricultural labourers were still malnourished: 'The agricultural worker was, in fact, the worst fed of all workers in the nineteenth century' and subsisted on a meagre diet of cheese, bacon and some milk.[10] The agricultural labourer was also the worst housed. In 1888 R.E. Prothero wrote in his *Pioneers and Progress of English Farming* that farm labourers were 'herded together in cottages which, by their imperfect arrangements, violated every sanitary law, generated all kinds of disease, and rendered modesty an unimaginable thing'.

We now view the cities of Victorian Britain with horror, yet labourers were attracted from the country into the cities by the higher wages. In 1900 A.L. Bowley wrote his *Wages in the United Kingdom in the 19th Century* to explain that northern agricultural workers received double the wages of those in the south because the demand for labour from the mills and factories was so intense. The Agricultural and Industrial Revolutions, by facilitating their move into the towns, thus bettered people materially. The revolutions may have driven people to drink, but at least they provided them with the money with which to buy it. In *Lark Rise to Candleford* (1945), her memoirs of rural life in Oxfordshire at the end of the nineteenth century, Flora Thompson revealed how, of an evening, village labourers were as desperate as city dwellers for a drink, but being so poor they had to nurse a single half pint of beer the whole evening long.

While Britain was industrializing it was still a poor country (though the richest in the world), and its Industrial Revolution was therefore easy to misrepresent, especially as urban industrial slums were then a novel horror. Its greatest critics were, of course, Karl Marx and Friedrich Engels, yet as David Felix chronicled in his 1983 book *Marx as Politician*,[11] Marx and Engels simply misrepresented the facts, lying systematically and portraying the working class as getting poorer when it was getting richer. So, to quote but one of his many lies, in his 1864 *Inaugural Address and Provisional Rules of the International*

Working Mens' Association, Marx claimed that Gladstone had said in his Budget speech of 17 April 1863 that, 'This intoxicating augmentation of wealth and power is entirely confined to classes of property,' whereas, as Hansard (the parliamentary record) confirms, Gladstone had rejoiced in the opposite.

The Industrial Revolution had its faults: it generated losers as well as winners, and it wage-enslaved once-free spirits. But it did enrich them materially, it did generate great public goods like the sewers and Public Health Acts, and, by creating an economy of self-sustaining growth, it eventually liberated humanity from its millennia of enslavement to poverty. And it was done with no significant government funding for science.

THE PRIVATE FUNDING OF VICTORIAN SCIENCE

As the Industrial Revolution developed, however, when purely technological development had been exhausted, so an interesting phenomenon emerged: the British increasingly supported science, but privately. Industrial Revolutionary Britain, under laissez faire, was to create magnificent science – witness such names as Davy, Faraday, Maxwell, Dalton, Kelvin, Darwin, Huxley, Lyell, Cavendish and Parsons.

There were at least four separate sources of funding for science under the free market in Britain: hobby science, endowed university science, industrial science, and industrially or fee-funded university science.

Hobby science

This term is borrowed from the hobby farmers, the rich men who improved their farms. There were many hobby scientists, some of

THE INDUSTRIAL REVOLUTION

whom were stunningly good. Few scientists are remembered after their deaths but, of those that are, three were British hobbyists: Henry Cavendish, the third Earl of Rosse and Charles Darwin.

Henry Cavendish (1731–1810), the grandson of two dukes, Devonshire and Kent, was for a time both the richest man in England and also England's greatest physicist. He identified hydrogen as an element, he characterized the atmosphere, he synthesized water from its elements, and he determined the density of the earth. He lived in Clapham, and for his delicate torsion balance experiments he rose at four in the morning, when vibration from horse-drawn traffic was minimal.

Being so rich, Cavendish could afford his little eccentricities. In the words of Lord Brougham he 'probably uttered fewer words in the course of his life than any man, not excepting the monks of La Trappe'. He was scruffy, he had no friends or lovers, and he spoke only to Fellows of the Royal Society – and then only on Thursday nights when a select few were invited to dine. He dismissed any servant he encountered around the house and, other than on Thursdays, his dinner would be served via a device that spared him the sight of his butler.

William Parsons, third Earl of Rosse (1800–67), was a great astronomer. By 1827 he was improving the casting of specula for reflecting telescopes and experimenting with fluid lenses. In 1845, at a cost of £30,000, he built his great reflecting telescope or Leviathan, which at 58 feet long, 72 inches in diameter and 4 tons in weight, was the largest telescope in the world until it was superseded in 1917 by the 100 inch reflector at Mount Wilson, California (which was also built by private philanthropists). Through it he discovered, among other new things, spiral nebulae. He was President of the Royal Society between 1848 and 1854.

Charles Darwin (1809–82) was also privately rich, having both inherited and married money. The five years he spent on HMS *Beagle* were unpaid.

Those three men were but the most prominent of thousands of hobby scientists. As the increasing wealth and low taxes of industrial Britain bred increasing numbers of rich people, so a remarkable number of them embraced science. Indeed gentlefolk once took to science as they took to the theatre, and during the nineteenth century Londoners attended the science lectures at the Royal Institution as they might have attended the opera. To cope with the press of fashionable carriages Albemarle Street, where the Royal Institution sits, became one of the world's first one-way streets.

Privately endowed university science

The importance and romance of research has long prompted private donors to endow academic science. Henry Lucas, for example, the MP for Cambridge in 1640, endowed the Lucasian Professorship of Mathematics at Cambridge to which Isaac Newton was elected in 1669 (Babbage was elected in 1823; Stephen Hawkins is the current holder). In 1619 Sir Henry Savile, previously Elizabeth I's tutor in Greek and mathematics, endowed the Savilian Professorships of Astronomy and Geometry at Oxford, to one of which Christopher Wren was elected in 1661. Sir Thomas Gresham, the founder of the Royal Exchange, also created Gresham College (1597) in London, to whose professorships were elected such men as Robert Hooke of Hooke's law and, again, Christopher Wren.

By the nineteenth century the rate of private endowment had vastly accelerated, and professorships, departments and entire universities were privately created throughout Britain. The universities included Manchester (1851), Newcastle upon Tyne (1852), Birmingham (1900), Liverpool (1903), Leeds (1904) and Sheffield (1905). (Those are the dates of the Royal Charters, showing how the universities, though often rooted in older institutions, were nonetheless created after the Industrial Revolution.) The greatest privately endowed academic science, though, emerged from London. Some of this came from the University of London (1836), whose constituent colleges such as Imperial College became world famous, but overshadowing even the university was that great privately endowed research foundation, the Royal Institution (1799).

The Royal Institution was created by Count Rumford, an interesting American. Born in 1753 as Benjamin Thompson at Woburn, Massachussetts, he joined the colonial army, serving in a New Hampshire regiment, but in 1776, proving himself an honest patriot loyal to his oath to his king (as well as a bit of a spy in the Benedict Arnold mould), he fled the revolting American colonies for England. There, he turned to military science, and his experiments on gunpowder prompted his election as a fellow of the Royal Society in 1779. He returned to New York in 1782 as a lieutenant colonel commanding a British regiment, but at the war's end he was unemployed, so in 1784 he entered the service of the king of Bavaria.

As Bavarian minister for war, Thompson established a cannon foundry where he studied the heat generated by the boring of cannon, disproving the caloric theory of heat (that heat is a substance). Francis Bacon in his *Novum Organum* (1620) had anticipated the modern, kinetic, theory that 'heat itself, its essence and quiddity, is motion and nothing else,' but by Thompson's day even a great chemist like Lavoisier could suppose heat to be a substance. Thompson destroyed Lavoisier's theory, restored Bacon's, and to complete his triumph even married Mme Lavoisier.

Thompson then reformed the Bavarian army, established a military academy, created a poor law, founded an institute for plant and animal breeding (which introduced the potato to those unenlightened parts) and laid out the English Garden in Munich (where Unity Mitford was to shoot herself in 1939 when her adored Adolf Hitler invaded Poland). Thompson was consequently made a Count of the Holy Roman Empire (taking the title of Rumford from a New Hampshire town), and by 1796 he was running Bavaria as president of the council of regency, the king having fled the approaching French and Austrians.

On his return to London, Rumford argued that British industry needed more scientific research – research that should be freely available. Rumford disapproved of patents: 'I desire only that the whole world should profit by [my discoveries] without preventing others from using [them] with equal freedom.' But to Rumford the provision of public goods was not a matter for government but for private philanthropy, and he wrote: 'We must make benevolence fashionable.' He succeeded. Within a year of its foundation in 1799 the Royal Institution had raised no less than £11,047. The money came from individual subscriptions, with membership costing from 50 guineas for founders' life membership to 2 guineas a year for annual subscriptions. And Rumford was himself a philanthropist, who left most of his estate to Harvard University to found the Rumford Professorship. Earlier he established the Rumford medal for the Royal Society (on condition he was the first winner).

The Royal Institution was not unique as an independently funded laboratory that published science freely. Its first lecturer, Dr Thomas Garnett, was lured from Anderson's Institution in Glasgow, which was a similar laboratory; and the Royal Institution's first giant, Humphry Davy (1778–1829), who was appointed lecturer in 1801, was lured from yet another independent research laboratory that published science freely, the famous Pneumatic Institution at Clifton, Bristol, where he had discovered the anaesthetic effects of nitrous oxide or laughing gas.

At the Royal Institution Davy was to discover six new elements (potassium, sodium, barium, strontium, calcium and magnesium), and in 1806 he delivered his seminal lecture 'On Some Chemical Agencies of Electricity'. On marrying a rich woman in 1813, Davy resigned his paid employment but continued to research as a hobby scientist, inventing the safety lamp in 1815 and becoming the president of the Royal Society in 1820.

Davy's great pupil, Michael Faraday (1791–1867), worked all his life at the Royal Institution. He was such a great scientist that a mere list of his achievements is awesome: let us just mention that in 1831 he discovered electromagnetic induction, the process by which electricity can be generated by rotating a coil in a magnetic field, which has underlain the commercial generation of electricity ever since. The Royal Institution has continued to foster great scientists, including James Dewar (of the vacuum flask) and four Nobel laureates (Lord Rayleigh, Sir William and Sir Lawrence Bragg and Lord George Porter). All supported by private philanthropy.

Industrial science

As industries developed, so the technology required for further innovation became more and more complex. Progressively, therefore, industrialists became scientists or scientists became industrialists. Typical was Charles Parsons, the youngest son of the third Earl of Rosse of the telescope, who invented the steam turbine and who built his celebrated engineering works at Heaton, Newcastle upon Tyne. The Parsons Works fostered a considerable research laboratory, and such laboratories spread through the technological industries of Britain. By 1900, for example, there were some 200–250 graduate chemists alone employed in British industry.[12]

(Parsons did a naughty thing during a Review of the Fleet. To celebrate Queen Victoria's Diamond Jubilee in 1897 the Royal Navy assembled 173 warships off Spithead to be inspected by the Prince of Wales. His Royal Highness was to sail in state down the vast lanes of warships in the royal yacht, followed by a flotilla carrying the colonial premiers, the Diplomatic Corps, the House of Lords and the House of Commons. Parsons, trying to interest the Admiralty in his turbines, stole the show by roaring around the fleet in *Turbinia*, his turbine-powered yacht, which at 2,000 horse power and a top speed of 34 knots was faster than anything the Navy could send in chase. It was, in fact, the fastest boat in the world. *The Times* wrote: 'Her speed was simply astonishing, but its manifestation was accompanied by a mighty rushing sound and by a stream of flame from her funnel at least as long as the funnel itself.' The Admiralty promptly ordered two.)

University science funded by industry

As industrial research became more scientific, so industrialists would commission or endow academic research. But the real impact from industry came through its demand for trained scientists. In the early days, men like Newcomen, Trevithick, Stephenson and Watt could train themselves. Others, like the younger Stephenson or Charles Parsons, were trained by their fathers. Indeed, industrial research dynasties were born: after David Mushet (1772–1847) discovered how to prepare steel from bar iron, his son Robert (1811–91) discovered how to produce cast steel.

But with the growing complexity of industrial research, industry increasingly needed to recruit trained scientists; it could no longer depend on self- or family-taught researchers. Industry, therefore, created a demand that the universities expanded to satisfy. The economics were simple: industry offered premium salaries against which employees would refund their university fees (just as American medical students today pay large fees knowing that their subsequent earnings will more than pay them off). Academic science flourished in consequence because, enriched by their fees, the universities competed to employ the best qualified teachers, who were necessarily the best researchers. To sustain those researchers' reputations, therefore, the universities had to find the money to sponsor their research.

The consequences

Under the influence of these four separate private sources of finance, Britain's research base grew dramatically. The historian Roy Macleod has calculated that Britain's science base, as determined by the numbers of scientists, doubled every twelve to twenty years during the nineteenth century.[13] The membership of the metropolitan science societies, for example, most of whose members were amateurs or hobby scientists, doubled from about 5,000 in 1850 to 10,000 in 1870. The ranks of the professional scientists had an even shorter doubling time of around twelve years, the numbers of university lecturers in science rising, for example, from sixty in 1850 to 400 in 1900.

And by 1914, according to *Who's Who in Science*, there were 1,600 leading scientists who had found 'congenial employments' where unhindered research was encouraged and funded.

WHY DID THE BRITISH FUND SCIENCE UNDER LAISSEZ-FAIRE?

So, why did the British fund science under laissez-faire? The industrialists were clearly motivated by commercial opportunities, and their in-house laboratories were simply profit-seeking enterprises. Their commissioning of university science was similarly profit-seeking, and though academic research was apparently more risky for them (university research cannot be kept secret, so competitors can learn of it), they nonetheless benefited from trading those risks against their access to larger, university-wide research programmes than they could have funded in-house.

The interesting donors were the philanthropists. Why did they do it? Conventional economists assume humans solely to be profit-maximizers, so they cannot account for philanthropists. But the biologists know that we humans evolved to seek, above all, the rewards of the social hierarchy: what people *really* want is status. Some people are sufficiently sexy, witty, articulate, creative, artistic, literary, musical, adept or scientific to achieve status directly via careers in modelling, acting, politics, writing, art, music, dancing, sport or science, but those people are rare. Most of us, therefore, seek status via money.

While they are making money a person will indeed behave as a strict profit-maximizer, eschewing philanthropy. But once that person has made their money, they may well barter it for the currency that really matters, approbation. Consider Andrew Carnegie. While he was building up United Steel, Carnegie behaved as a mean, penny-pinching accountant; as president of United Steel he'd have no more paid for commercially irrelevant fundamental research into metallurgy than he would have bought drinks for the house. As he once said: 'There's no money in pioneering.'

But in 1901 Carnegie sold his businesses to J.P. Morgan for $492 million, upon which he mutated into one of the most exhaustive philanthropists in history. As he explained in his 1889 auto-biographical *Gospel of Wealth*, 'he who dies rich dies disgraced', and before he died in 1919 he had given away over $350 million. Among his endowments were 2,811 free public libraries, 7,689 church organs (though he was agnostic) and, in 1905, $10 million for the Carnegie Foundation. That foundation was for long the most generous funder of science on the planet.

Or consider John Rockefeller of Standard Oil and various railways. On his way to the top he was not always a nice man, but once he'd made his money he gave away no less than $530 million, endowing institutions as varied as Chicago University (1891; the home of Milton

Friedman) and the Rockefeller Foundation (1913; among other crucial discoveries it funded Oswald Avery's discovery that DNA conveyed genetic information and Howard Florey's development of penicillin).

Yet Carnegie and Rockefeller were only marching to an old English drum. Consider the wealthy Thomas Plume (1630–1704). He built the Plumian Library in his home town of Maldon, Essex, as one of the world's first free lending libraries; he endowed £1,200 to establish a college of weaving in Maldon; and he endowed £1,902 to create the Plumian Professorship of Astronomy and Experimental Philosophy at Cambridge University (of which a recent holder was Fred Hoyle). Other British philanthropists include Samuel Courtauld, the textile magnate, who in his *Ideals and Industry* (1949) explained why he had created his Courtauld Institutes of Fine Arts and Biochemistry, and Henry Wellcome, whose trust remains the largest charitable foundation in the world and which funded much of the sequencing of the human genome.

Yet the Brits themselves were only marching to an older drum. It was Petrarch who, observing the remarkable philanthropy that emerged out of the thriving market economy of fourteenth-century Italy (witness Brunelleschi's Foundlings' Hospital in Florence) observed that 'aristocrats are not born but made'. Such generosity continues today, and it is intense in the US (one of the freest economies in the world), whose citizens in 2002 gave away $186 billion (1.8 per cent of their GDP as opposed to the 0.8 per cent, say, that the British give of their GDP).

Today the Howard Hughes Foundation (assets $11 billion) supports science generously; and both Bill Hewlett ($4 billion) and Dave Packard ($9 billion) endowed scientific research. And note that the Bill and Melinda Gates Foundation owned assets of $30 billion *before* Warren Buffett made his own extraordinary gift of some $35 billion. Today there are over fifty American philanthropic foundations whose assets exceed $1 billion (including the Lilly Endowment, $10 billion; and the Ford Foundation, $9 billion) and many of those are devoted to research.

Friedrich Nietzsche wrote that humans 'need to give', and the modern science of neuroeconomics confirms that philanthropy is hard-wired. There is a technique known as functional Magnetic Resonance Imaging (fMRI), which allows scientists to monitor the activity of those parts of the brain that signal reward (the bits that light up when we eat, drink, and partake of recreational drugs and sex). Using it, a team led by Ulrich Mayr, a psychologist at the University of Oregon, studied the responses of subjects when they were invited to give money to a charity that helps the poor. Mayr found that, on their so doing, their brain reward centres lit up: people like giving money to good causes.[14]

Mayr entitled his paper 'Neural Responses to Taxation and Voluntary Giving Reveal Motives for Charitable Donations' because he found that people also responded positively to being taxed: if people's money is taken away from them mandatorily to a good cause, people's brains light up, but their brains light up *less* than when they give money voluntarily. Mayr therefore worries that taxation simply crowds out or displaces voluntary giving. Regardless, Mayr's paper shows that philanthropy is biologically determined and that it is a myth that only governments will succour good causes.

In another recent paper studying the charitable donations of subjects in laboratory settings, the psychologist Vladas Griskevicius of Arizona State University showed that when people believed that their deeds would be noticed (especially by the opposite sex) they became more philanthropic.[15] Consequently, much philanthropy is no mystery because a public good is *public*. After Pliny the Younger (61-113) built a free library and supported a school for poor children, he described in his letters the 'popular favour and applause' he enjoyed. All the great cultures have institutionalized philanthropy: Islam has the *wakf* and *zakat* systems, Jewish philanthropy is famous (the Hebrew words for justice and charity are the same, and Israelis give away the highest percentage of GDP in the world), and the Hindus have the concept of *ahinsa*. Every human culture acknowledges munificence, such that it can even become pathological. The Native Americans of the Northwest Pacific coast, for example, lived by 'potlatch', by which men ruined themselves financially as they competed for social status by providing their peers with the most generous presents and the grandest feasts.

Potlatch, being competitive, shows how philanthropy, in its demand for approbation, can be a form of bullying and may therefore provoke rejection. When in 1835 the Englishman James Smithson left the then-huge sum of £100,000 to the federal government in Washington, D.C., to found his institution, many congressmen did not want it. William C. Preston, a representative from South Carolina, denounced the gift as 'too cheap'. If Congress accepted it, he warned, 'every whippersnapper vagabond . . . might think it proper to have his name distinguished in the same way'. Senator John C. Calhoun maintained that it was 'beneath the dignity of the United States to receive presents of this kind from anyone'. Only in 1846 did the federal government finally accept the money.[16]

And when in 1937 Andrew Mellon offered $15 million for a national gallery, to be built across the Mall from the Smithsonian, to house the gift of his Old Master paintings (then valued at $80 million) suspicious politicians resented the donation. Congressman Wright Patman wrote to President F.D. Roosevelt that: 'The precedent is a very bad one. If we allow Mellon this privilege, Hearst and Morgan will come in next with an offer just as attractive.' It took five hours of rancorous debate for the Senate to accept the offer.[17]

So we see how, contrary to the economists' models, rich men and women will compete to provide public goods. So in November 2003, for example, National Public Radio in America received $200 million from Joan Kroc, whose husband had founded McDonald's restaurants. With a few more endowments like that, NPR will be autonomous in perpetuity, unlike the UK's BBC which, dependent on government patronage, will still be in hock to politicians and will still be sacking its chairmen and directors general when, as in 2004, they offend government (a shocking episode). Equally, in 2002 Ruth Lilly in America left $100 million to *Poetry* magazine and the Poetry Foundation, while we Brits are still stuck with the government's sclerotic Arts Council. (The UK Arts Council was created in 1939 to boost civilian morale, mimicking the wartime creation of the government science agencies).

And consider lifesaving. The British possess the safest seas in the world: if you had to choose a spot to risk drowning, you should choose the waters off the UK. Why? Because the Royal National Lifeboat Institution (RNLI) provides a superlative lifesaving service. In the seas off most countries, a person in peril has to rely on government-funded coastguards, but in the UK that person has the RNLI. The RNLI maintains no fewer than 320 lifeboats and it rescues some 6,000 people a year. Since its foundation in 1824 it has saved 135,000 lives.

The RNLI is the twelth largest charity in the UK, with an income of £128 million annually. It charges nothing of anyone. Instead, its every penny is provided voluntarily, by myriad donations from millions of people. The RNLI accepts *no* government money. If a person were to suggest to a conventional economist that a free lifesaving service could be provided by a market society, that person would be ridiculed – yet there stands the RNLI, a perpetual reproach to the economists' perceptions of public goods. (Holland and Germany have similar charitable bodies.)

Here is the fundamental point: the market depends on mutual trust, otherwise contracts will never be observed. A successful market society, therefore, fosters trust. By extrapolation – and the functional MRI data confirms this – the human emotions that animate the market will also animate philanthropy. The philanthropic impulse is thus the extension of the commercial one, and needs no more explanation within a successful market economy than does the observance of contracts.

Indeed, the quickest way of destroying philanthropy is for the state to support public goods. Consider a fascinating episode from the history of the RNLI. After thirty years of existence the RNLI got into financial difficulties, so in 1834 it accepted £2,000 a year in government subsidies. Those rose over time, but the RNLI found that, with every pound it raised from the government, the rate of voluntary donations fell by *more* than a pound; private individuals would not support a state-supported institution. Moreover, the bureaucracy and regulations that the government imposed on the RNLI damaged the actual service of lifesaving. So in 1869 the RNLI cut loose from its subsidies: that was a brave act, which people predicted would bankrupt it; but instead – once the RNLI was seen to be solely dependent on private philanthropy – the philanthropy burgeoned.

The state in the west now provides 'free' science, 'free' education, 'free' health and 'free' welfare out of taxes, so the private philanthropic provision of those services has been crowded out; but where the state ignores public goods (such as the federal government in Washington, D.C., which largely ignores art galleries) then private providers flourish, to provide a service that is *superior* to that supplied by governments elsewhere; US art galleries are the best in the world.

* * *

The Industrial Revolution took place in Britain because, after 1688, people felt safe to invest in profit-making activities, and, contrary to

Bacon's thesis, people invested in R&D. Moreover, they sometimes did so as philanthropists as well as profit-maximizers. Science, therefore, must be different from what Bacon said it was. What might it be? Here is a clue.

We saw earlier that David and Robert Mushet were a research dynasty, with David discovering how to prepare steel from bar iron, and Robert Mushet discovering how to produce cast steel. But Robert wanted to monopolize his discoveries. As L.T.C. Rolt recounted in his 1970 book *Victorian Engineering*:

> In 1862 Robert Mushet established the Titanic Steel Works in Dean Forest. This was a small crucible steel-making plant where, in great secrecy, Robert Mushet carried out his alloy steel experiments . . . and R. Mushet's Special Steel or 'RMS' soon became famous in machine shops on both sides of the Atlantic. Mushet took the most extraordinary cloak-and-dagger precautions to keep his RMS formula secret. The ingredients were always referred to by cyphers and were ordered through intermediaries. The mixing of the ingredients was carried out in the seclusion of the Forest by Mushet himself and a few trusted men.

But Robert's secretiveness backfired, and he died poor and embittered. In that failure we will see that science does not consist of discreet nuggets of potentially public knowledge that Bacon described; it actually consists of nuggets of tacit knowledge that can be traded. I shall return to this point shortly.

11

Much Economic History since 1870 is Only Convergence

'A mere copier of nature can never produce anything great.'
Sir Joshua Reynolds, Discourse to the Students of the Royal
Academy, 14 December 1770
'Those who do not want to imitate anything, produce nothing.'
Salvador Dali, *Dali by Dali*, 1970

God, during the nineteenth century, was British. One little offshore nation, Great Britain, dominated the world. The citizens of that small, foggy island controlled the greatest empire in history: London ruled a quarter of the human race and a fifth of the surface of the globe, including India, Burma, much of Africa, Canada, Australia, New Zealand, much of the Caribbean, chunks of South America and China, and bits around the Mediterranean, Antarctica and Polynesia. The Royal Navy policed the oceans – Britannia ruled the waves – and the British army preserved world peace. *Pax Britannica*. Underpinning that vast enterprise was the British economy, also the greatest the world had ever seen. British products inspired awe from San Francisco to Irkutzk, from Helsinki to Cape Town. Birmingham was the workshop of the world.

Countries are competitive, and when they become Top Nation they like to boast. So in 1848 a civil servant called Henry Cole, later Sir Henry or 'King' Cole, conceived of an international industrial exhibition. He proposed that products from factories all over the

SEX, SCIENCE AND PROFITS

world be collected, to celebrate Britain's supreme contribution to industry. Cole was already a well-known figure if only because, in 1843, he had sent the world's first Christmas card, and if only because he had improved the design of the teapot, yet not everyone liked him. Queen Victoria disparaged his 'rough offhand manner', but Prince Albert supported him with a joke: 'If we want the exhibition to *steam* ahead we must have *Cole.*'

Not everyone wanted an international exhibition. Colonel John Sibthorpe, MP for Lincoln, prayed in the House of Commons that the exhibition might collapse into bankruptcy: 'This unwholesome castle of glass is attracting to our shores a riff-raff of swindling foreigners, robbing and murdering us by licence of the Queen's German husband.' The government, moreover, refused to help financially: in those days governments followed Robert Peel's dictum that:- 'Of all the vulgar arts of government, that of solving every difficulty which might arise by thrusting the hand into the public purse is the most delusory and contemptible.'[1]

So Albert and Cole first set themselves the task of fundraising, soon collecting over £100,000 from individual donations (no commercial or corporate sponsorship was accepted). They then had to find an architect. They envisaged an enormous building, one that covered 20 acres, but they could not wait years for it to be conventionally built, so they needed a revolutionary technique of construction. Fortunately, Joseph Paxton, the Duke of Devonshire's gardener, had perfected the art of prefabrication.

Paxton (1801–65) had built a number of large greenhouses on the ducal estates at Chatsworth, and he had trained the local foundries to cast metal casements to be assembled on site. Using this technique, Paxton built so fast that in a matter of months he had erected in Hyde Park in central London a huge greenhouse, nicknamed the Crystal Palace, to house the exhibition. In 1851, only three years after Cole's memorandum, the queen opened the Great Exhibition of the Industry of all Nations.

The Great Exhibition was a fantastic success. It attracted 13,937 exhibitors and, during its 5½ months, over 6 million visitors (when Britain's population was 21 million), and the international jury awarded British artefacts most of the prizes. The takings were £423,792, and the profits £186,000. Even the Grand Opening Procession was a success, having been led by the Duke of Wellington, who had earlier made a crucial contribution by suggesting an ecological solution to the problem of the pigeons (the Crystal Palace had been rapidly colonized by large numbers of pigeons, and their droppings caused much distress). The obvious solution, sniper fire, was impossible, but during a meeting of the Privy Council the duke had suggested 'Sparrowhawks, Ma'am'.

A century later, in 1951, another international exhibition, the Festival of Britain, was held in London to celebrate Britain's contributions to industry and design. The Festival proved another triumph. It attracted 10,250,000 paying visitors over five months, it transformed the South Bank of the Thames into today's arts centre, and it endowed the Royal Festival Hall and other cultural monuments whose dividends London still enjoys.

How dismayingly those triumphs of 1851 and 1951 compare with Britain's recent Millennium Dome! To celebrate the 2000th anniversary of the birth of Christ, four UK politicians, John Major, Michael Heseltine, Tony Blair and Peter Mandleson, commissioned a celebratory Dome to be erected in London. It was, in the words of one its begetters, Peter Mandleson: 'to knock peoples' socks off.' In the words of another of its begetters, Tony Blair, its opening ceremony was 'to hold the attention of the entire world . . . in the most important building in the world'.

It was a disaster. The opening ceremony, held at midnight on New

Year's Eve, was a shambles, with invitees queuing for over three hours in the cold and the dark while the really important people, including the Rt Hon Anthony Blair PC, QC, MP swept past, Kremlin-style, to their seats. Afterwards, exposing the snobbery that animates New Labour, Lord Falconer, the responsible government minister and one of Tony's cronies, apologized 'both to VIPs and to ordinary people' for the discomfort.

The Dome, moreover, attracted only half the numbers planned, 6 million visitors instead of 12 million, so it received only half as many people monthly as the 1851 and 1951 exhibitions (a quarter of Disneyland's, incidentally). The Dome's tickets at £20 ($30) were expensive compared to the shilling (£3 in today's money) for the cheapest tickets for the 1851 exhibition, and the Dome's finances were tainted by the Hinduja scandal, ministerial resignations and losses of nearly £1 billion.

For much of the twentieth century Britain was in decline, and the fiasco of the Dome, built to mark the end of that century, seems only to have highlighted that decline. Indeed, everything about the government's millennial celebrations was tacky. Norman Foster's 'Cool Britannia' Millennium Bridge had to be closed after three days, and the 'River of Fire' firework display was a damp squib. And, to national astonishment, a Frenchman trained by an American company, Pierre-Yves Gerbeau of Disney ('Monsieur Gerbil') had to be imported to keep the Dome running. The only millennial success was the London Eye, which was, of course, a private initiative.

Friedrich Nietzsche once described Britain as the 'nurse of history', yet for a century the British seem to have been nursing themselves into post-industrial decline. Why do lead countries decline? One answer people always offer when they are trying to explain national rates of economic growth is government funding (or its lack) for science. So as early as 1852, just one year after the triumph of the Crystal Palace,

Lyon Playfair (1819–98), a prominent chemist, published his *Industrial Instruction on the Continent* to claim that European industry would soon overtake Britain's unless the British government invested more in science. And Playfair reinforced his warning after the frightening Paris Exhibition of 1867.

The 1867 Paris Exhibition was similar to the 1851 London Great Exhibition except that it was larger, covering 41 as opposed to 20 acres. The outcome, however, was very different. In 1851 the international jury had awarded most of the prizes to Britain, but in 1867 Britain won only ten of the ninety, being overtaken by France, Belgium, Germany and the USA. Whereupon Playfair wrote an open letter arguing yet again that British industry was in decline because the government did not fund enough science.[2] This letter so inflamed public opinion that in 1868 the government set up a Select Committee to examine its claims. The Select Committee dismissed them:

> Although the pressure of foreign competition, where it exists, is considered by some witnesses to be partly of a superior scientific attainment of foreign manufacturers, yet the general result of the evidence proves that it is to be attributed mainly to their artistic taste, to fashion, to lower wages, to the absence of trade disputes abroad, and to the greater readiness with which handicraftsmen abroad, in some trades, adapt themselves to new requirements.
>
> *Select Committee on Scientific Instruction*, 1868

But the Select Committee's dismissal of Playfair has not dented the almost unanimous verdict of historians that Britain entered long-term technological decline during the mid- to late nineteenth century. Consider these comments:

The worst symptoms of Britain's industrial ills . . . was the extent to which she was no longer in the van of technical change; instead, even the best of her enterprises were usually being dragged in the wake of foreign precursors, like children being jerked along by important adults.

David Landes, *The Unbound Prometheus*, 1969

Britain's technological defeat was first alarmingly revealed . . . in 1867.

Correlli Barnett, *Audit of War*, 1986,

Britain's relatively poor economic performance [1870–1914] can be attributed largely to the failure of the British entrepreneur to respond to the challenge of changed conditions.

D.H. Aldcroft, 'The Entrepreneur and the British Economy 1870–1914', *Economic History Reviews*, 1964, 17:113–34

At the turn of the century, a new international pattern of technology was emerging. Britain was in decline, Germany was in the ascendant.

T. Williams, *The Triumph of Invention*, 1987

Britain, hitherto industrially supreme, had been very obviously outclassed in the Paris international exhibition . . . most notably by Germany.

Margaret Gowing (Professor of the History of Science at Oxford University), 'An Old and Intimate Relationship', Spencer Lecture, Oxford, 1982

Germany developed one of the best technical and education systems in the world. This . . . was one of the main factors in Germany overtaking Britain in the latter half of the nineteenth century.

Chris Freeman and Luc Soete, *The Economics of Industrial Innovation*, 1997

But did the 1867 international exhibition really expose British decline? Let me first dismiss the suggestion that the different number of prizes won by Britain in 1851 and 1867 means anything. The numbers of British entries were very different. Of the 13,937 exhibitors in 1851, half were either British (6,861 exhibitors) or from the British Empire (520 exhibitors): the competition was unbalanced towards the host country. A similar imbalance occurred in 1867 but tilted, of course, towards France. Half the exhibitors were French, and the Belgians and Germans also found it easy to exhibit, while Americans made a big effort to be represented. There were relatively few British entries, and so relatively few British prizes.

Second, we must distinguish between relative and absolute decline. Let us simplify and represent economic strength as oranges. Let us pretend that Britain in 1851 was so rich that it possessed five oranges, while the USA, Germany, France and Belgium each owned only one orange. Now let us pretend that, by 1867, all five countries had done equally well and that each had acquired a further orange. Britain would now possess six oranges, but the USA, Germany, France and Belgium would own two apiece.

Two consequences flow: first, Britain no longer emerges as richer than everyone else combined, because Britain's six oranges are outnumbered by its competitors' eight; Britain, therefore, is in relative decline. Second, its competitors have grown at higher percentage rates: to double your oranges from one to two is to achieve 100 per cent growth, to increase your number of oranges from five to six is to achieve only 20 per cent growth. Yet, in absolute numbers of oranges, everyone has grown equally, by one. On the orange analogy, therefore, it is absurd to have expected Britain to have won the majority of the prizes in 1867. The year 1851 hit the moment when Britain had five oranges to its competitors' one, but the situation could not have lasted – nor should it have. In a just world, everybody should have six oranges.

The third reason for dismissing the prizes won at international competitions as indicators of national wealth is that they are not, in fact, indicators of national wealth; they are indicators of national technological achievements – which may never translate into wealth. If an international exhibition had been held in 1957, say, just after the USSR had launched its first sputnik, an artificial orbiting space satellite, the USSR would have been showered in prizes. Japan, on the other hand, which by 1957 had aspired to little more than copying other people's technology, would have won few prizes. Yet whose economy has since flourished? In 2000 Russia's GDP per capita was $1,740. Japan's, for all its problems, was $38,000, twenty times greater.

To judge economic achievements, therefore, we must turn away from anecdotes about international exhibitions; we must turn to the systematic economic historians. After all, Gustave Eiffel, whose Eiffel Tower was built for the 1889 International Exhibition in Paris, was sentenced in 1893 to two years' imprisonment for fraud over the building of the Panama Canal, but could anyone extrapolate from that episode to argue that the entire French political class would today be tainted by corruption?

THE PATTERN OF ECONOMIC GROWTH UNDER MARKET CAPITALISM

Between the years 500 and 1500 per capita wealth in Europe did not increase. Thanks to the new technologies, there had been a consistent increase in absolute wealth of about 0.1 per cent a year, but that had been absorbed by an equivalent population growth rate. From 1500 onwards, however, coinciding with the stirrings of the Commercial Revolution, rates of wealth increase started to outstrip rates of population increase, and standards of living started slowly to creep up. Italy led the way, but after that country was invaded the economic leadership of Europe passed to the Netherlands. But they, too, were invaded, leaving Britain – safe behind the English Channel – to enjoy the fruits of 1688. Figure 11.1 illustrates Britain's GDP per capita since 1700. (The legends to the figures are on page 438.)

Figure 11.1 GDP per capita UK

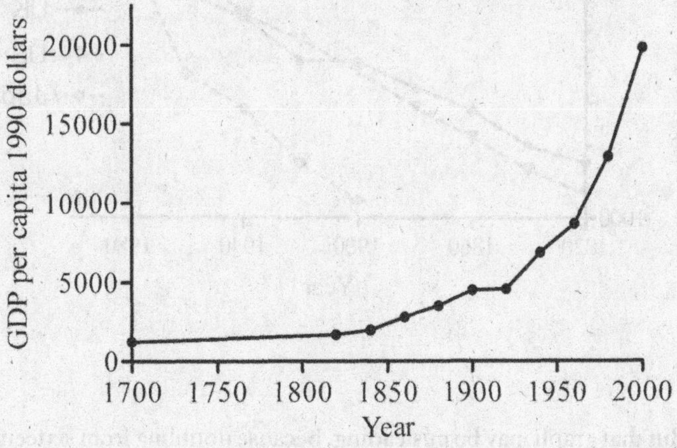

Figure 11.1 shows that the British have got richer, but it doesn't show the rates at which they've enriched themselves: did Britain grow faster during the 1720s, 1820s or 1920s? A conventional figure like Figure 11.1 does not easily reveal *rates* of growth. Consider a breeding pair of rabbits, Benjamin and Cottontail. If Benjamin and Cottontail produce four offspring, each pair of which then produces four more, the numbers of rabbits in each generation will be 2, 4, 8, 16, 32, and so

on. That regular doubling will produce a rapidly rising curve rather like that in Figure 11.1.

Figure 11.2 GDP per capita UK, US and Japan

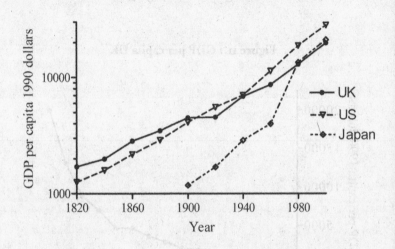

But that graph may be misleading, because doubling from sixteen to thirty-two rabbits produces a bigger rise than from four to eight, yet the actual *rate* of growth is the same, namely a doubling. To look at rates of growth we need a logarithmic graph, as in Figure 11.2 (above) If the rate of rabbit reproduction is constant, irrespective of the numbers of rabbits breeding, then the curve becomes a straight line. And Figure 11.2 confirms that the rates of per capita growth in Britain have, indeed, remained roughly constant since 1820 at a bit below 2 per cent a year, having doubled every twenty-five or so years for nearly two centuries. We can therefore conclude that between 1700 and 1820

Britain's rate of economic growth increased gradually until, around 1820, it hit the peak rate, which it has subsequently maintained.

The essentially constant nature of sustained growth among economic lead countries is illustrated by the USA in Figure 11.2. The US economy has grown at an astonishingly linear rate for nearly 200 years, as have, incidentally, the economies of countries such as Canada and Switzerland. If, like the USA, Canada and Switzerland, lead capitalist countries can avoid being invaded, they will grow smoothly, at least in the long term. In the short term, countries' economic growth rates are variable and are determined by transient factors such as the exchange rate, consumer spending or business confidence, but in the long term the transient factors cancel themselves out and countries hold to their underlying rates of growth.

Figure 11.2, moreover, illustrates another phenomenon, that of 'convergence'. The countries that emerge into capitalism late enjoy higher rates of growth than do the pioneers, because the latecomers are converging on the rich economies. Thus in Figure 11.2 Japan starts to grow later than Britain, but then overtakes that country. France shows a similar pattern. The Baconians would have us believe that these countries accelerated growth can be attributed to its government funding of science, but since the French government was funding science long before its economy started to converge on Britains's, we can dismiss that argument. Moreover, Britain has now overtaken France, and that British acceleration followed the cuts in government science budgets that Shirley Williams, Margaret Thatcher and other politicians initiated (see later).

But the Baconians were once vocal about Japan's economy which, as Figure 11.2 shows, shot ahead thanks to MITI, the Ministry for International Trade and Industry. Really? This story is worth pursuing because it reveals how pervasive can be the myths of government support for science.

JAPAN

For centuries Japan was feudal, economically stagnant and isolationist, being closed to foreigners at pain of death. But in 1853 the US Navy sent in Commodore Matthew Perry and his Black Ships to force the free entry of American goods and traders. Humiliated, and fearing that the Yanks were even then preparing their smallpox-infested blankets and incubating concepts of Manifest Japanese Destiny, samurai from Satsuma and Chosu concluded that the isolationist policy of the Tokugawa shoguns had failed, and in 1868 they seized the imperial palace in Kyoto to restore the Emperor Meiji to real power and to set Japan to copy the West and industrialize.

But in 1868 Japan was so ignorant of Western practices that the emperor and his government needed to be taught even how to copy. So at vast expense they imported Western teachers, and between 1868 and 1872 no less than 40 per cent of the Japanese government's budget was spent on salaries to foreign experts.[3] Copying had become the major arm of state policy.

One typical foreign expert was Henry Dyer, an engineer from Glasgow University, who in 1873 was invited to create the Tokyo Polytechnic, later to become the Faculty of Engineering of Tokyo University. To lure Dyer, the Japanese paid him four times what he would have earned in Britain. Indeed, the men who hired him earned less: in 1873, a Japanese cabinet minister received 500 yen a month, but, as the principal of the Tokyo Polytechnic Dyer was paid 660 yen a month.

Japan copied eclectically. It copied its educational, parliamentary and legal institutions largely from Germany, which is why Japan's schoolboys still wear black, flat-peaked caps Bismarck-style, which is why its parliament is called the Diet, and which is why it has Roman not common law. But Japan copied British traffic law (it drives on the left) and, more importantly, it copied British and American economic policy, namely laissez-faire. Having germinated the market, after 1884 the Japanese government understood that it runs best unfettered and left it alone.

This is ironic because until recently it was fashionable to affirm that Japan prospered because of MITI's direction of the economy. Yet since 1884 Japan's government has actually seen itself as the facilitator of industry, anxious to tax it and its customers as lightly as possible. Japan remains one of the three least-taxed of the major industrialized countries, the other two being the USA and Switzerland, as is shown by the employment statistics: 1 in 4 workers is employed by the state in France, one in six in the UK, one in seven in both the US and Germany, and only one in eleven in Japan[4].

And MITI, far from being a brilliant leader of industry, has been wrong so often that the Japanese themselves now acknowledge that they have grown despite, not because of it. MITI had opposed the development of the very areas where Japan has been most successful: cars, electronics and cameras. Yet MITI poured vast funds into wasteful projects. Until recently, for example, MITI provided Japan with a huge overcapacity in steel – no less than three times the national requirement – which was a mistake of perverted genius because Japan possesses few natural resources other than raw fish, so it had to import everything (the iron ore, the coal, the gas, the limestone and the oil) to make its unwanted steel.

MITI also invested in giant, loss-making, fifth-generation super computers at the moment that the market opened for the small personal computer; and MITI's attempts at dominating the world's pharmaceutical and telecommunications industries have failed. Indeed, MITI's interventions into the Japanese economy since 1955, across hundreds of different companies in all major sectors of the economy, have harmed virtually every industry and company they touched.[5]

Contrary to myth, much Japanese technology has been incompetent. Consider Japan's space programme, which is years behind schedule. In 1999, typically, Japan's National Space Development Agency aborted that year's H-11 launch by exploding it a few seconds after take-off; in 2000, another typical year, Japan's Institute of Space and Astronautical Science lost its Astro-E.

Or consider Japan's nuclear record: its fifty-five nuclear plants are plagued by accidents. In 1999 a leak from the Tokaimura nuclear power station, which was operated by untrained workers with Chernobyl-style incompetence, irradiated sixty-nine people at lethal levels and drove the 310,000 citizens of Tokai to barricade themselves at home behind sealed doors and windows as the radioactive cloud engulfed their town. A leak in 2004 at the Mihama plant killed four people, and a fire broke out in 2006 at the Ohi plant. And consider the *Mutsu*, Japan's only nuclear-powered ship. Built in 1975, it has leaked radiation since its maiden voyage. For over sixteen years it was in dock while engineers repaired it. In May 1990 it was finally relaunched for sea trials, only for them to be aborted when the crew was exposed to frightening amounts of radiation. The project has now been cancelled, at a final cost of $600 million.

Japan's biological sciences, too, are oddly backward. For example, Japan licensed the contraceptive pill only thirty years after Europe or the USA because, as Carl Djerassi, the inventor of the pill, noted on 15 February 1996 in *Nature*, Japan has long used abortion as its primary means of birth control (over a million abortions a year), and the $400 million-worth abortion industry had successfully lobbied MPs for three decades to obstruct the pill. And organ transplants, though now legal, are still rare. Interestingly, though, Regaine/Rogaine (the baldness treatment) and Viagra were licensed the very moment they were discovered, baldness and impotence being issues for the geriatric males of the Diet.

These words are not written to knock Japan – far from it, it is an impressive country – they are written to emphasize that Japan grew not because its government funded science (because it did not, see later) nor because MITI was clever (it was not) but because it had, long before 1868, under the social contracts of its type of feudalism, developed the cultural underpinning required of a disciplined market economy and its people were already widely literate.[6] All it had to do after 1868 was to copy Britain, Germany, the USA and other lead countries, and its path was so straightforward that it could afford the occasional mistake, no matter how bad. (After 1970, once it had attained the per capita wealth of other lead nations, Japan should have copied their corporate governance and financial disciplines; instead, it fuelled artificial growth rates by shameless commercial practices, but that is another story.)

The important message for this book is that during the post-war period, when Japan was growing fast by convergence under laissez-faire, and when the Japanese government funded a *smaller* share of its national academic science and R&D than any other OECD country, the pundits nonetheless turned it into a story of Japan flourishing thanks to MITI's direction and funding of companies' science and strategies. It seems almost impossible to persuade pundits to abandon their belief in science as a public good, and if an economy grows fast too many pundits will – in the face of the actual facts – find an 'explanation' in the government funding of science.

GERMANY

Germany is another 'other country' that we were once told grew miraculously because its governments funded science. Thus we were told that Germany overtook Britain economically during the nineteenth century because of the German government's investment in education and science. So, for example, the chemist Edward Frankland (1825–99), a pupil of Lyon Playfair's, claimed that Germany was publishing six times as many scientific papers as Britain. But, as the historian Roy Macleod has noted, no one has ever confirmed Frankland's improbable claims.[7]

Equally, 'Rhenish capitalism' (Rhenish means of the Rhine) has also been widely praised. Unlike the UK/US model, Rhenish capitalism fosters cartels and minimizes competition. In the 1890 Sherman Act, for example, the Americans had outlawed industrial cartels, but in 1897 the German Supreme Court declared price-fixing and market-sharing cartels to be not only legal but morally elevated because, by minimizing 'selfish' or 'wasteful' competition, they embodied a shared concern for the common good. And the three leading *Grossbanken*, Deutsche Bank, Dresdner Bank and Commerzbank, took shares in the companies in which they invested, thus sparing them from competing with each other (each bank holding shares in all the potential competitors). Then, in 1879, tariffs were introduced. Thus did the Germans abandon free trade and thus (we are told) did the Kaiser's Germany overtake Victoria's Britain.

But did it? If we ignore the anecdotes and examine the quantitative data of the economic historians, we discover that during the nineteenth century German economic growth rates were no higher than Britain's, at 1.3 per cent a year.[8] This is a devastating fact, and it represented failure for Germany, because Germany started the nineteenth century with a GDP per capita only 75 per cent of Britain's, and by 1914 Germany's GDP per capita was . . . 75 per cent of Britain's. *Germany had failed to converge*, which was one reason it lost the First World War. Similarly, as late as 1939 Germany was still only 75 per cent as industrialized or as rich as the UK. That was one reason it lost the Second World War.

In addition, because during the nineteenth century other poor countries, including Russia and Italy, did converge on Britain, Germany grew increasingly poor, relatively – as the Germans themselves well knew. The Kaiser, notoriously, told the banker Max Warburg on 21 June 1914 that he was contemplating an early war because Germany's economy was falling behind and he was losing his window for victory.

During the nineteenth and early twentieth centuries the Germans were not only poor, they sometimes literally starved. A series of terrible harvests from 1845 precipitated mass malnutrition (and the 1848 continental revolutions). The urban British during the Industrial Revolution may not have been especially healthy, but they were healthier than the Germans. As late as 1914 a German infant was almost twice as likely as a British infant to die in infancy, and British life expectancy at birth was one to three years longer than in Germany.[9] No wonder Lieutenant Ludwig Wittgenstein of the Austrian Artillery (and later of the philosophy faculty at Cambridge) wrote in his diary for 25 October 1914: 'The English – the best race in the world – *cannot* lose.'

The Germans were so poor that between 1841 and 1900 no fewer than 4.9 million of them emigrated. Many left for the USA, but many also settled in the UK, where they flourished (Siemens, Mond/Melchett, Halle, Goschen, Casel, Schroder). Contrary to myth, the Germans found Britain more entrepreneurial than their own country. This is what Jacob Behrens the émigré banker wrote in 1832:

I took a liking for Britain, especially because it presented a picture totally different from retrogressive Germany . . . Not only did I feel myself a man amongst men, but the times were great.

Quoted in P. Panayi, *German Immigrants in Britain during the 19th Century*, Berg Publishers, Oxford, 1995

IRON AND STEEL

During the nineteenth century, therefore, the British, in the absence of government funding for science, kept well ahead of the Germans, despite their huge subsidies. It is only by anecdote or misrepresentation that some historians have woven their myths of British decline. Let me here expose the myths behind two anecdotes, the rise of the steel and chemicals industries in Germany.

Until 1880 iron and steel was Britain's industry. In 1880 Britain produced 46 per cent of the world's pig iron and 36 per cent of the world's steel – more than the USA and double the amount produced by Germany – and it was the world's dominant exporter. But thereafter British production levelled out, and by 1913 Britain was the world's largest importer of iron and steel. An industry that once accounted for 12 per cent of British GNP had fallen to only 6 per cent and sinking.

Criticism has been poured on to British education and research for this failure. In his *Audit of War* the Baconian Correlli Barnett wrote:

Before 1884 there was no university department of metallurgy anywhere in Britain, no university research, [but] the Germans looked from the very start to organized science and technology, to thorough training at every level, as the necessary instrument of future industrial success. The forging of this sophisticated and elaborate tool began half a century and more before the resulting tempered blade began cutting into world markets: the Berlin Technical Institute for engineers dated from 1821; technical high schools at Karlruhe from 1825, at Dresden from 1828, at Stuttgart from 1829. By around 1870 there were already some 3500 students in German technical high schools. Since the early 1820s German universities had likewise been developing formidable teaching departments in chemistry, metallurgy, physics and engineering. They established research laboratories as well, to act as pathfinders to future technological leadership.

It all sounds jolly impressive, and, indeed, some important people were jolly impressed. In his *Diaries* for 17 October 1988 Alan Clark, the government minister, wrote: 'I was interested and gratified to hear her [Mrs Thatcher] pass a comment showing she had read *Audit of War*.'

But let us look at the facts. The three great technical advances in steelmaking during the second half of the nineteenth century were Bessemer's invention of the converter in 1856, which made mass production possible, Siemens's invention of the open-hearth furnace in 1866, and Thomas's invention in 1879 of the 'basic' process, which enabled steel to be extracted from phosphoric ores. And the curious thing about all three inventions was that they were made in Britain.

Sir Henry Bessemer (1813–98) was an amateur, untrained in science, who wrote in his *Autobiography* that his knowledge of metallurgy was: 'very limited, and consisted only of such facts as an engineer must necessarily observe in the foundry or smith's shop'. Wilhelm Siemens (1823–83) was, admittedly, German in origin; he had trained as a metallurgist in Germany, and he was assisted by Pierre Martin (1824–1915), a French metallurgist, but significantly, Siemens moved to Britain, where he created his branch of the company and where he did his research, finding the entrepreneurial spirit of the UK more conducive to industrial research. The discovery in 1879 by

Sidney Thomas (1850–85) of the 'basic process' was, perhaps, the most consequential, because it rendered most ores suitable for the Siemens–Martin open-hearth furnace – and Thomas was another amateur, a hobby scientist who worked as a clerk in a police court in Wales and who experimented in chemistry at night.

Thus the suggestions that Britain's steel industry was overtaken by Germany's because German science was superior are false. And Britain made the advances for the same reasons it always had; not because it possessed vast, gleaming, government-funded laboratories of white-coated scientists, but because it possessed men like Bessemer, Siemens, Martin and Thomas, who were close to the market.

If British metallurgical science was so good, why did the industry decline? Relative decline domestically was, of course, inevitable. Britain could not continue to devote 12 per cent of GNP to iron and steel after 1871 or the island would now be knee-deep in rolling mills. Relative decline vis à vis foreign steelmaking was also inevitable; other countries were bound to develop their own steel industries. Thomas's discovery, moreover, preferentially benefited the Germans and other competitors because continental ore, unlike Britain's, is largely phosphoric.

The USA, Germany and France, moreover, raised tariffs against British steel; and Britain's competitors subsidized their industries in ways that the British did not. Consider training. In Britain workers were trained as apprentices on the job or in the 700 or so mechanics' institutes that the free market had generated, privately, between the 1820s and the 1840s to offer technical instruction.[10] (The myth is that only the state will provide education, but nearly 200 years ago, under the market, Britain had already produced 700 independent mechanics' institutes to educate the workforce.) But the institutes' costs were met, ultimately, by industry because the artisans' fees, or their loans to meet the fees, were translated into higher wages by which the artisans could repay the costs of their education. In Germany, however, steel workers were trained in technical schools, whose costs were born by the State. Since the costs of training in technological industries can, at 15 per cent or more of annual turnover, dwarf profits, the consequence was that the German steel industry was heavily subsidized, and the British one was not. The cumulative effect of this unilateral subsidy was to undercut British steel.

Moreover, the German government paid companies to export steel – and it fostered a domestic cartel to further subsidize exports. The British *Iron and Coal Traders Review* for 24 December 1909 concluded that: 'without its vast system of syndication – its almost military-like production and distribution methods – and the organized fostering of export trade by countries, the German iron and steel industries could hardly have obtained their present status.'

The British response to this foreign pressure was rational because it was market driven: it gradually closed down its industry. The purpose of economic activity, as Adam Smith had noted, is to consume – production is only a means to an end – and if German taxpayers wanted to subsidize steel exports to Britain, that was Britain's good fortune. For British steelmakers the decline of their industry was grim, but for British consumers steel had become cheap, so ultimately the British economy benefited.

But in subsidizing steel German taxpayers were not so rational, because they had an ulterior motive. They had a heavy schedule of invasions ahead of them (Denmark 1864, Austria 1866, France 1870, Belgium *et al.* 1914, Czechoslovakia 1938, Poland *et al* 1939), and for those they needed steel. In his famous *Blut und Eisen* (Blood and Iron) speech to the Prussian House of Deputies of 28 January 1886, Bismarck had stated: 'Place in the hands of the King of Prussia the strongest possible military power, then he will be able to carry out the policy you wish; this policy . . . can be carried out only through blood and iron.'

Well, Germany got its iron, but at the cost of taxes, cartels and tariffs. And Germany got its blood too; much good it did it.

CHEMICALS

Let me consider one more of Britain's mythological 'failures', that of the chemical industry. The conventional view was stated by Professor Denis Noble of Oxford University: 'The first major government intervention in science in the UK, in 1913, was to save a nation approaching a state of war with a country on which we had become almost totally dependent for chemical products.'[10]

By 1913 Germany accounted for 24 per cent of the world's chemical output, and the USA, with its much larger population, 34 per cent, but Britain accounted for only 11 per cent. The outbreak of war in 1914 did, indeed, find Britain alarmingly exposed chemically. Britain had grown dependent on Germany for many chemicals, not least the khaki dye for its soldiers' uniforms, and 1914–18 witnessed desperate shifts as British chemists and industrialists struggled to provide *matériel*.

Yet criticisms of the British chemical industry are, curiously, hard to sustain. First, the chemical industry grew faster between 1881 and 1911 than any other industry in Britain except for public utilities, and by 1911 it was employing 2.7 per cent of the manufacturing labour force.[12] Britain did well in detergents, paints, coal tar intermediates and explosives (explosives are useful in war). The British chemicals industry, therefore, cannot be criticized for not growing, only for not growing as fast as Germany's. This must be a general criticism because nobody – neither the USA, nor France, nor Belgium, nor the Netherlands – expanded into chemicals, especially organic chemicals, like Germany. Moreover, Britain soon learned its strategic lesson. During the First World War it created the British Dyestuffs Corporation, which in 1926 combined with Nobel Industries, Brunner Mond and the United Alkali Company to create ICI, the chemicals giant that by the late 1930s was employing nearly 500 research scientists and many of whose managers were scientifically trained. Chemically, Britain was fully prepared for war in 1939.

British chemical companies made mistakes, of course. Specialist historians complain that Britain's United Alkali Producers retained for too long the Leblanc process for synthesizing sodium carbonate, when it should have shifted more of its production into its licensed Solvay process after 1897. But it can be difficult to scrap profitable technology for untried technology. Consider Britain's electronic industry. This, too, is often criticized, but for the opposite offence, namely for an overenthusiasm for new technology. The major British electronic company, GZ de Ferranti, was obsessed by new technology, and Ferranti always installed it, even if the older technology was more profitable or more reliable.

However, these individual mistakes are small when put into context. United Alkali Producers' mistake over the Leblanc process cost it around £1.9 million,[13] yet the company still prospered, and if one episode of a forgone £1.9 million can be described as one of the worst examples of British 'entrepreneurial failure' before 1914, there could not have been much wrong with the British entrepreneurialism of the day.

The Germans excelled in organic chemistry only for the same reasons they excelled in iron and steel – the government poured vast funds into technical schools, universities and scientific research. But again, the fundamental discovery was made by a Briton in Britain, when William Perkin (1838–1907) in London synthesized mauve, the first of the aniline dyes, in 1856. For millennia dyes had been extracted from natural sources (150,000 marine snails had to be sacrificed for each Roman emperor's purple toga), yet natural dyes were not only expensive but they generally faded fast. Perkin's discovery of aniline dyes, by opening a huge new market for chemistry, transformed it. But Perkin made his discovery in the classic British amateur mode. He was only eighteen years old when he made it, experimenting in an attic at home.

Perkin also had a day job, as an assistant at the Royal College of Chemistry. The director of the College, however, was a German, Wilhelm Hoffman, who studied Perkin's science carefully, and when he returned home in 1865 he used government money to train a generation of young scientists in Perkin's science, to fuel Germany's dominance in dyes, fertilizers and drugs. But by their taxes in support of their chemicals industry, the Germans paid opportunity costs elsewhere: it's worth re-emphasising that Germany during the nineteenth century failed to converge on Britain.

GERMANY'S SO-CALLED ECONOMIC MIRACLE

Economic growth in Germany during the nineteenth century was largely achieved, as in Japan, not by science but by copying. As in Japan, Germany imported foreign experts. The Hamburg and Bergedorf railway, for example, was designed in 1838 by an Englishman, William Lindley. British labourers built the actual railway too, because the Germans in 1838 were still pre-industrial in their culture, being lazy and incompetent.

People tell anecdotes of British amateurishness, but here are some anecdotes of German amateurishness. Consider the motor car. The first practical car was produced by Gottlieb Daimler and Karl Benz. Yet the key inventor was neither Daimler nor Benz but Daimler's boss,

Nicolaus Otto. It was Otto who, in 1876, created the first practical internal combustion engine, on which all subsequent automobile development has depended. Yet Otto was an amateur in the British mould. No *technische hochschule* graduate he: instead, he was a travelling salesman who, happening to read a newspaper article on Jean-Joseph Lenoir's 1859 inefficient 2-stroke engine, decided on the instant, in a moment of destiny, to create an efficient internal combustion engine (4-stroke was the key).

Even when great German discoveries were made by professional scientists, they could be made in endearingly British-type ways. Consider nitrocellulose, which is the basis of modern gunpowder and a key chemical. It was discovered in 1845 by the German chemist Christian Schönbein (1799–1868), who discovered it after accidentally spilling some nitric acid at home on the kitchen table, where his wife had absolutely forbidden him to experiment with chemicals. To hide his guilt, Schönbein immediately rushed to wipe up the nitric acid with his wife's cotton apron, which happened to be hanging over the stove to dry, when – whoosh! – there was a flash, a puff of smoke and no apron. Christian may have discovered nitrocellulose but, boy, was he in trouble with Frau Schönbein.

Another great German chemist who got into trouble with his wife was Adolf von Baeyer (1835–1917) who, in 1864, discovered barbiturates. Having family money, he was a hobby scientist, researching unpaid at Berlin University, having earlier trained in Friedrich Kekulé's private laboratory. And on discovering his new family of chemicals, he naughtily named them 'barbiturates' after his lover, Barbara. (Barbie, as in doll, was also named after a Barbara, hence Jacqueline Susann's *Valley of the Dolls*.)

Kekulé (1829–96), in whose private laboratory Baeyer trained, was the most romantic of scientists, making all his great discoveries (such as the ring structure of benzene) while dreaming:

During my stay in Ghent I lived in elegant bachelor quarters . . . I was sitting writing my textbook but the work did not progress; my thoughts were elsewhere. I turned to the fire and dozed. The atoms gambolled before my eyes. My mental eye could distinguish large structures of manifold conformation, long rows all twining and twisting in snake-like formation. But look! What was that? One of the snakes had seized hold of its own tail!

Kekulé describing his 1864 discovery of the ring structure of benzene[14]

Kekulé's later discoveries were paid for by the state, because after his early successes he was employed by the University of Bonn, but the early discoveries that made his name were funded by family money.

Another innovator whose work was funded by family money was Count Ferdinand von Zeppelin (1838–1917) of airship fame. By 1908, however, he had exhausted most of his inheritance on R&D, and when his LZ 4 crashed that year he announced that he would have to withdraw from any further airship work. Whereupon the German people, in an outburst of spontaneous mass generosity, collected no fewer than 6 million marks by which he could continue his research. It is a shame that his work culminated in the launching of Zeppelin bombing raids on London between 1914 and 1917, but that is a different story.

We rarely learn of how Otto or Schönbein or Baeyer or Kekulé or Zeppelin funded their discoveries because no pressure group publicizes their stories: white-coated, government-funded scientists like to pretend that research can be performed only by white-coated government-funded scientists such as themselves. Instead, we often hear of Germany's economic miracle, but where is this famous miracle? When did Germany enjoy a standard of living as high as the US's or Switzerland's or even Luxembourg's? Germany did, admittedly, overtake the UK during the late 1960s, but so did fifteen other industrialized nations. The post-war period was one of British relative decline, not of German miracles. Paradoxically, moreover, Germany started to converge on Britain and the other lead economies only after 1948 when, ironically, it abandoned its Rhenish capitalism.

After 1948 a group of German economists who called themselves Ordo-liberalists captured the control of the economy. Under Ordo-liberalism (as in 'Order-liberalism') markets are subject to the rule of law but not to the government: the Ordo-liberalists believed that autonomous markets are morally as well as economically superior to both cartels and socialism because they liberate individuals to make free choices.

The giant Ordo-liberalist was Ludwig Erhard (1897–1977), the economics minister. In office from 1948, he acted as Milton Friedman before the day, establishing an independent central bank, progressively eliminating government controls on the economy, introducing the sound Deutschmark, resisting price controls or the inflationary printing of money, and pushing through the Anti-Cartel Law (1952–7). And during the 1950s he progressively liberalized trade, driving tariffs down from an average of 20 per cent at the beginning of the decade to 10 per cent by the end.

Erhard was opposed on all sides, but by the late 1950s everyone acknowledged his success, as Germany's economy forged ahead. Only latterly, as Germany has reverted to a more Rhenish capitalism, has its economy stuttered again.

* * *

For lead countries, long-term economic growth rates are limited to about 2–3 per cent per year because their growth is the consequence of innovation, and innovation is difficult. Converging countries, which are only copying, can grow much faster – until they have converged. But, despite the anecdotes, there is no systematic evidence that the government funding of science helps either lead or converging countries do any better than they would otherwise have done.

Because of the systematic failure of economic historians to show that the government funding of research has ever helped economic growth, the Baconians hop around to find another 'other country' whose government's example we have to follow. Once the other country was France, then Germany, then Japan, then the USSR (see next chapter) and now the USA (see the chapter after that). But each of these anecdotal examples always crumbles on examination.

12

Nationalizing American Knowledge

'The night they drove old Dixie down
And all the people were singin'
They went, "Na, na, na, na, na, na, na, na, na". '

Robbie Roberston, 'The Night they Drove Old Dixie
Down', 1969

America became the richest country in the world around 1890 without significant government money for science. Since 1940, however, the federal government has invested so generously in research as to have effectively nationalized much of it. Let us ask what that has achieved.

Science is mentioned in the American Constitution – 'The Congress shall have power . . . to promote the progress of science and the useful arts' (Article 1, Section 8) – but in practice during the eighteenth century the American government did little for science other than, in 1790, passing a Patent Act. During the early part of the nineteenth century a few specialized federal institutions were created, including the Library of Congress in 1800 and the Coast Survey in 1807, and though those institutions did research, none existed primarily for it; they were created to perform particular functions, and any research was secondary to their mission. Research for its own sake was not believed to be a responsibility of government, as was illustrated by the foundation of the Smithsonian Institution.

James Smithson (1765–1829) was the English scientist who in his will left over £100,000 (then $550,000): 'To found at Washington under the name of the Smithsonian Institution, an Establishment for the increase and diffusion of knowledge among men.' Smithson was the illegitimate son of the Duke of Northumberland, and he never forgave England the consequent social exclusion: 'My name shall live in the memory of man when the titles of the Northumberlands and the Percys are extinct and forgotten.' Having never visited America, he imagined it a haven of tolerance.

As we know, this gift was not welcome to a nation that did not believe that research was a responsibility of the federal government. John C. Calhoun, senator from South Carolina, warned that: 'We must look carefully at the extent of our own power. This Government is a trust, established by the states, with a specific capacity, education not included ... What are we to do with the money? There is no difficulty in that, it must be returned to the heirs.'[1] It was not until 1846 that Congress finally accepted Smithson's money, and then only to prevent any more being siphoned off by corrupt politicians in Arkansas, where, recklessly, it had been banked. Congress would not spend taxpayers' money on science, yet America's economy did not suffer because early America was a poor country that, sensibly, was growing by copying British technology.

The most famous North American industrial copier was Samuel Slater (1768–1835), the British emigrant who in 1789 built America's first factory, a cotton mill at Pawtucket, Rhode Island. Slater had been lured by the bounties that the US offered to skilled cotton workers, and he needed the incentives because he broke British law. In those days of Navigation Acts and of mercantilism, Britain prohibited the export of industrial designs. Indeed, the British would not even allow textile workers like Slater to emigrate. But Slater played dumb, pretending to be so stupid that the port officials in Britain thought America the very place for him, and he sailed to Rhode Island with the details of the latest British spinning machines hidden in his clever brain.

Much early American engineering was led by Brits. America's first major canal, the Middlesex Canal (1793–1803), which linked Boston with Middlesex County and southern New Hampshire, was largely the creation of the British engineer William Weston. And even after the Americans had weaned themselves off skilled British immigrants, they still copied British technology. Consider Andrew Carnegie. Carnegie had made steel in a small way during the American Civil War, but it was only on a visit to Britain in 1875 that he learned of Sir Henry Bessemer's famous process. On his return to the US Carnegie built his own Bessemer steelworks, and he soon captured 65 per cent of America's production, but his technology was British.

Moreover, the Americans trained themselves on the job. French engineers trained in their *académies*, the Germans in their *technische hochschulen*, but American engineers, like their British counterparts, trained on the job. Indeed, the Americans even gave their system a name, the Erie system, which was named for New York State's Erie Canal, which was built between 1817 and 1820 by three surveyors, Benjamin Wright, Canvass White and James Geddes. Initially, those three men knew little about canal construction, but they talked to men who had worked with British ex-pats, they examined existing canals, they read, they experimented, and as they built, they learned. And as they learned, they taught their assistants. These, in turn, rose to become chief engineers elsewhere, spreading the Erie system of on-the-job training all over America.

Adam Smith would have approved. He did not think that formal or government-funded education was necessarily better than any other kind: 'When a man has learned his lesson well, it surely can be of little importance where or from whom he learnt it.'[2]

THE CIVIL WAR

This story of technological laissez-faire, however, was sharply reversed on the outbreak of the American Civil War (1861–65). It was Alexander Bache (the great-grandson of Benjamin Franklin) who pushed Congress into funding research. Bache (1806–67), a tidal expert who was the superintendent of the Coast Survey, had long lobbied for an American academy of science, and in 1851 he had even created a club, the Lazzaroni, to foreshadow one, but it took Civil War to create a real Academy with real federal money and real federal prestige, on the French model.

The Civil War presented the federal government with urgent technical problems: ironclads needed to be built, gunnery improved, and steam engines converted to military use. Bache lobbied, and in 1863 Congress established the National Academy of Sciences with Bache as its first president. Consequently, the Academy established committees to advise government departments. On 8 May 1863, for example, a committee met to consider how to operate a magnetic compass in a ship built of iron, while another met to report on electroplating ships with copper to prevent corrosion and fouling in sea water. But much of the work of the Academy was shared with a near-identical body, the Permanent Commission of the Navy Department. The Commission was created a month earlier than the Academy, yet it possessed an overlapping membership and it did similar things. Why two bodies?

One of Bache's strongest supporters for an academy had been Joseph Henry (after whom the unit of conductance or henry was named), who was the secretary of the Smithsonian but who had assumed that Congress, because of its historical suspicion of science, would never create an academy. So, separately, he had lobbied Gideon Welles, the Secretary of the Navy, to create a Permanent

Commission by executive action. In the event the legislature also obliged, leading to the creation of both the Academy and the Commission. Fortunately, common sense prevailed and Bache, who chaired committees of both bodies, had fun writing to himself, as the chairman of one committee to another, asking himself for advice.

Bache needed his fun because his Academy offended many scientists. On 23 May 1863 *Scientific American* led the opposition, complaining that government support for science would lead to government control, which would impede the disinterested pursuit of truth; that Academy members were not to be paid, so the government would obtain its scientific advice for nothing, whereas before it had paid the consultancy fees of a learned profession; and that because members would not be paid, the Academy would be staffed largely by scientists who were already employed by the government rather than by those who earned their living in the marketplace. It would promote, therefore, the interests of government-employed scientists rather than those of science itself.

In the event, subsequent peacetime federal administrations soon lost interest in the Academy, and though they recognized it as their scientific adviser, successive administrations paid it no standing grant, meeting the expenses only of specific investigations. Bache, however, was content. He had wanted an academy, and he had used the Civil War to get one. As we shall see, the foundation of the National Academy of Sciences was characteristic. There will always be scientists who believe that science should be advanced with government funds, governments will resist the pressure, but, time again, war persuades them to create scientific institutions. Once created, institutions are hard to destroy because they mobilize political support. Scientific institutions are particularly resourceful: in war they explain how they can create weapons, in peace they explain how they can create wealth. And they rarely boast about their origins in war.

THE MORRILL ACT

The American Civil War also precipitated the first federal intervention into higher education, the Morrill Act of 1862, which created agricultural colleges across the nation. An earlier act had been opposed by defenders of the free market and of states' rights, and though in 1859 it had passed both Houses of Congress, it had been vetoed by President James Buchanan. But with the departure of the southerners the Morrill Act became law in 1862.

People, even today, celebrate the Morill Act, but why did America need federally funded agricultural colleges? The problem with American agriculture, even during the mid-nineteenth century, was overproduction. Since medieval times, as we know, farmers could grow huge amounts of food, and only by restrictions on the market could farming be made profitable: thus England had erected its first food tariffs or Corn Laws as early as the fourteenth century, and, to help alleviate the overproductivity, the British government had introduced subsidies for corn exports as early as 1689. In *The Review* of 1713 Daniel Defoe had urged the expansion of gin production because: 'Nothing is more certain than that the ordinary produce of corn is much greater than the numbers of our own people or cattle can consume.'

If those were England's agricultural problems, how much worse were America's with a thin population, near-limitless, high-quality virgin land and ever-improving transport! Yet agriculture was America's largest industry and, as its troubles worsened through overproductivity, so governments had to be seen to be doing something, however illogical (and Morrill's colleges were illogical because, as land-grant colleges, they were funded by the sale of federal land to farmers, the very people who were meant to be so poor as to need federal assistance). Yet the Morrill Act was itself only part of a greater economic malaise – mercantilism.

The Jeffersonian vision of a laissez-faire America, where taxes were minimal, federal power was minimal, and trade was free, had been challenged by Henry Clay; and under Clay's 'American system' tariffs and taxes were to be raised, and industries, including canals, roads, railways, agriculture and research, were to be subsidized. Clay's disciples in 1862 included Congressman Morrill and Abraham Lincoln. Indeed, in his earlier Morrill Tariff of 1860, Morrill had doubled tariffs from an average of 24 per cent (on a narrow range of items), to 47 per cent (on a much wider range of items) thus converting tariffs from money-raising taxes into blatant protectionism.

So the 1862 Act establishing agricultural colleges was simply another expression of Morrill's mercantilism, substituting government direction for the market. And mercantilism was damaging – and not just economically. The 1860 Morrill Tariff contributed to the Civil War because, by protecting the north's industries, it so raised the prices of manufactured goods as to enrage the south. That is why Clause 1 of the Confederate Constitution prohibited 'duties or taxes on importation from foreign nations'. Mercantilism had divided the nation.

FORMAL EDUCATION UNDER LAISSEZ-FAIRE

Other than in farming, federal intervention in science decelerated after the Civil War. Although various bureaux, including the Geological Survey, continued to be created, they represented little more than the rationalization of defined federal functions. Between 1900 and 1915, for example, the federal government made only four inquiries of the National Academy of Sciences. Yet under laissez-faire the American economy continued to flourish, as did its engineers. At the 1851 Great Exhibition in London, for example, Isaac Singer displayed his sewing machine and Cyrus McCormick and Obed Hussey displayed their reaping machines. In 1868 Michael Kelly of De Kalb, Illinois, invented barbed wire, in 1874 Christopher L. Sholes (1819–90) of Milwaukee invented the first effective typewriter, and in 1879 James Ritty of Dayton, Ohio, invented the cash register.

The American engineers and inventors of the day were, as we know, generally trained on the job, but the 1890s witnessed a revolution: formal engineering education suddenly accelerated. In the 1890s the numbers of engineering colleges suddenly expanded, the profession suddenly embraced formal qualifications, and by 1900 the majority of new engineers had been formally educated.[3] Why the volte-face?

Before 1890 the US had been a converging economy, but around 1890 the US overtook the UK in terms of GDP per capita. Since wealth is largely the consequence of technology, we know that around 1890 the US also overtook the UK technologically. Consequently, American engineers found themselves facing novel technological problems they could not solve simply by copying British inventions or by applying common sense – they needed a formal education, which the market supplied. There had long been a handful of engineering colleges in the US, including Norwich University (1820) and Rensselaer (1824), but after 1890 their numbers exploded. Before 1890 employers hired men they would train on the job but after 1890 employers would hire only college-trained engineers. And the market met the need because the new engineers, like American doctors today, commanded higher salaries to commute their college fees.

The shift under laissez-faire from self-taught engineering to formal academic study was illustrated by the career of Thomas Edison (1847–1931). Like many of the pioneers of the earlier British Industrial Revolution, Edison was a barely literate, uneducated artisan, and he had been expelled from school after only three months for being 'retarded'. So he learned on the job, working as a telegraph operator on the railways. Teaching himself to become an electrical researcher, Edison read Michael Faraday's *Experimental Researches in Electricity*, and as Edison read Faraday's book he realized he needed also to copy Faraday's research institute, the Royal Institution in London. So in 1876 he created his own at Menlo Park, New Jersey, and soon he was employing over twenty other researchers. Before his death Edison had filed no fewer than 1,069 patents in fields as diverse as ticker tapes, the carbon granule microphone, the phonograph and the electric light bulb.

But, as his technical ambitions widened, so Edison was led to probe ever deeper into pure research, discovering, *inter alia*, the 'Edison effect' – that is, that electricity flows only one way between a heated filament and an electrode. The Edison effect was to underpin the development of diodes, electric valves and the early electronic industry. So we see how the market leads the most commercial of men into the deepest of science. And Edison really was commercial. Consider the story of capital punishment by electrocution.

At vast expense, and by ingenuity and courage, Edison had pioneered the domestic supply of electricity, which he supplied as direct current (DC). But during the late 1880s Westinghouse commercialized Nikda Tesla's invention of alternating current (AC), the form we now use. AC is better than DC because it can be carried for longer distances. Alarmed by the commercial threat, Edison played dirty, claiming that AC was dangerous and proclaiming that 'just as certain as death, Westinghouse will kill a customer within six months'. To prove his claims, Edison publicly electrocuted dogs and cats with AC, using a small current to prolong the animals' distress so that some took over half an hour to die.

But Edison really needed a human sacrifice, which came in 1890 in New York when a William Kemmler was sentenced to death. Edison offered to perform the execution himself, not by the usual hanging, but by AC current. That would show how dangerous it was! Westinghouse, concerned, paid for Kemmler's appeal, but Edison appeared for the prosecution, and the execution proceeded. On the day, during the first passage of electricity, the wretched Kemmler writhed and twisted as the too-low current passed through his body and as the witnesses gagged on the smell of cooking human flesh. After the current was turned off Kemmler was still alive, so he was subjected to a further 15-minute period of roasting spasms until, with parts of his body bursting into flames, he finally managed to die.

Edison had made his point: Young Sparky sparked. Yet Edison and Tesla and Westinghouse and Menlo Park had also made a deeper point, showing that the market could produce the science as well as the technology, higher education and training that an advanced economy needed. But, yet again, this scientific laissez-faire was to be abandoned with America's reimmersion in war.

THE GREAT WAR

The parallel between the scientific responses to the Fist World War and the Civil War – with the creation of the National Research Council under George Ellery Hale matching the creation of the National Academy of Sciences under Bache – was remarkable, because Hale was a reincarnation of Bache.

George Ellery Hale (1868–1938), an astronomer, was the most influential member of the National Academy of Sciences, and when President Woodrow Wilson delivered his ultimatum to Germany on the 18 April 1916, he sprang into action. On the very next day Hale called an emergency meeting of the National Academy, which resolved to help the president, and by June the president had agreed to the formation of a National Research Council (NRC).

Under Hale's chairmanship the scientists of the NRC developed techniques for detecting submarines, they developed poison gases, and they developed IQ tests, but Hale, a Baconian, always had his deeper plan. As he wrote later in a private letter: 'While most of my own experience as chairman was during the war, the plans I always had in mind looked forward to work under peace conditions.' On 27 March 1918 he wrote to President Wilson to explain how a peacetime National Research Council would 'stimulate pure and applied research for the national welfare'.

Hale recruited industrialists and politicians to lobby for him, and on 10 May 1918 President Wilson signed an executive order making the NRC permanent. Hale had hoped for an Act from Congress, but that was then under the control of the Democrats, the party of the people, who distrusted the public funding of science as corporate welfare. Democrats saw science as a capitalist activity: if science was of benefit to industrialists, let them pay for it.

So the 1920s witnessed a reversal to pre-war days, with the federal government funding only applied research that was integral to its agencies' particular missions. Not that America suffered, because the private funding for science soared. Philanthropists poured money into university science (the most prominent gift being the $25 million the Bambergers gave in 1929 to the Institute of Advanced Study in Princeton, Einstein's home from 1932 until his death in 1955). Industry, too, poured money into research: between 1920 and 1931 the number of industrial research labs rose from fewer than 300 to some 1,600, and the number of research employees rose from around 6,000 to around 30,000.[4] Annual industrial expenditure on research soared from $25 million in 1920 to $120 million in 1931. No wonder the Twenties, powered by electrification, cars and planes, Roared.

The 1930s, however, were dreadful – not least for federal science

– because in response to the Great Depression the federal government cut back its spending. Between 1932 and 1934 the Department of Agriculture, the federal government's largest science agency, cut its research budgets from $21.5million to $16.5 million, while a purer research agency like the Bureau of Standards more than halved its budgets, from $3.9 million to $1.8 million. Who inflicted these savage cuts? The Democrats, under President Franklin Delano Roosevelt.

Roosevelt's predecessor, the business-friendly Republican Herbert Hoover, had protected the federal research budgets, explaining in a 1932 speech that 'progress is due to scientific research', but Roosevelt and the Democrats believed that improvements in productivity had caused the Depression and were aggravating it by throwing people out of work. In 1935 *Science* collected statements from ordinary people: 'There should be a slowing up of research in order that there may be time to discover, not new things, but the meaning of things already discovered', 'The physicist and the chemist seem to be travelling so fast as not to heed or care where or how or why they are going. Nor do they heed or care what misapplications are made of their discoveries.' Even Einstein believed that R&D created unemployment by making workers redundant.[5]

We now understand the Depression to have been caused not by improvements in productivity but by financial policy errors, including the US's bizarre banking laws, the Federal Reserve Bank's incompetence, the Hawley–Smoot tariff and so on, but the Democrats believed that if Hoover supported science, that elite activity of rich college boys looking for new machines to displace hands, then science had to be a bad thing. It was Roosevelt's Secretary for Agriculture, Henry Wallace, who eventually reversed that sentiment. Wallace believed that economic recovery required an increase, not a decrease, in federal expenditure to stimulate aggregate demand, and he encouraged Roosevelt to expand the federal research budgets as part of the New Deal. Consequently, scientists received federal support in every state in the Union.

Today, scientists agitate for federal money, but in Wallace's day, odd though that might now seem, many scientists resented his funds. Wallace complained to *Science* in 1934 that 'in the past, most scientists and engineers were trained in laissez-faire classical economics,' so they were 'a handicap rather than a help' to his expansion plans. But Wallace (who was, bizarrely for a future vice president, a Marxist) complained that 'the scientists have turned loose upon the world new productive power without regard to the social implications', which was his code for saying that governments should control science, if necessary by funding it.

But the scientists did not want to be controlled, if only because the market provided for science so handsomely. Between 1931 and 1938 the number of industrial labs in the United States rose from 1,600 to 2,200, their research personnel rose from 30,000 to 40,000, and their annual expenditure rose from $120 million to $175 million. The rise in industrial research monies between 1931 and 1938 may have been smaller than between 1920 and 1931, but it more than compensated for the cuts in the federal science budgets.[6]

Table 12.1 (page 256) illustrates the happy situation. In 1940 research and development (R&D) in the USA was dominated by the private sector, which spent $265 million. The government spent only $81.1 million, and this was concentrated on two sectors, defence and agriculture, neither of which was important to economic growth. Moreover, basic or pure science was well supplied with some $31 million a year by the universities' and foundations' own endowments, and that figure was further boosted to over $45 million a year because industry spends some 7 per cent of its R&D budget on basic science. Clearly, therefore, there was no market failure in American science under laissez-faire before 1940, which was why the scientists were reluctant for the federal government to intrude.

The rise in private expenditure on science during the Depression surprises some analysts, who suppose that expenditure on science rises only if national income rises. But that dogma holds only over the long term; in the short term industrialists will be empirical, and if – as in America after 1929 – they find scientists' salaries to be very low, they'll invest in human capital (cheap) rather than in physical capital (expensive). And during the first half of the twentieth century, which was largely a time of laissez-faire in research, US productivity doubled, as was shown by Robert Solow the Economics Nobel laureate.[5]

VANNEVAR BUSH

If there is a constant in US research, it is that there is always a powerful person waiting in the wings, itching to impose under his leadership a federally funded science policy on the people of America. That person, though, has always needed a war. The Civil War gave Alexander Bache his National Academy of Sciences, the Great War gave George Ellery Hale his National Research Council – and the Second World War was to give Vannevar Bush (1890–1974) the National Defense Research Committee, the Office of Scientific Research and Development *and* the Joint Research and Development Board.

Vannevar Bush (no relation to George or Kate) was the president of the Carnegie Institution, the private body that was then the USA's largest science funding agency. Yet Bush did not believe that science should be left to the voluntary sector:

> We have no national policy for science. There is no body within the government charged with formulating or executing a national science policy.
>
> Vannevar Bush, *Science: The Endless Frontier*, 1945

Before the war, no one listened to Bush, but fortunately Japan started to invade its neighbours. By 1940 war between America and Japan looked so likely that Bush could lobby the White House, and on 27 June 1940 President Roosevelt created by executive order the National Defense Research Committee (chairman: Vannevar Bush). The NDRC promptly moved to integrate the country's research with the needs of war, forming subcommittees (ordnance, chemistry and explosives, communications and transportation, instruments and controls) on which academics, industrialists and service officers sat.

In June 1941 the NDRC was adopted by Congress as the OSRD or Office of Scientific Research and Development (chairman: Vannevar Bush) with access to vast funds, and by 1945 federal expenditure on R&D had risen twenty-fold to $1.591 billion, the increase being almost entirely attributable to defence and atomic research. Yet the OSRD, though successful in delivering the Manhattan and other projects, was only a wartime measure, and with the threat of peace came the perennial threat of demobbing. It was to help convert his OSRD into a national research foundation that, at war's end in 1945, Bush wrote his famous book, *Science: The Endless Frontier*, to argue that:

> Basic research leads to new knowledge. It provides scientific capital. It creates the fund from which the practical applications of knowledge must be drawn.

Consequently, in July 1945 Senator Warren G. Magnusson, one of Bush's Baconian allies, introduced a bill proposing a national research foundation. But he was opposed by Senator Kilgore, who countered with a bill of his own.

AN UNEXPECTED DISAGREEMENT

Harley Kilgore, Democratic Senator of West Virginia, agreed with Bush that the US needed a national research foundation, but whereas Bush wanted a meritocratic foundation that would be run by scientists for scientists for the purpose of creating the best science, Kilgore believed in a New Deal-type foundation like the WPA (Works Progress Administration), that would employ scientists equitably. So Bush wanted to distribute grants according to scientific merit, but Kilgore wanted grants to be distributed equally across the states. Bush wanted a foundation run by scientists, but Kilgore wanted one run by federal officials. Bush wanted the scientists to retain patent rights, Kilgore wanted the government to retain them. And Bush denied that the social sciences were proper sciences, whereas Kilgore wanted to support them.

For five years Washington was split on party lines. The Democrats, who distrusted scientists and industrialists, supported Kilgore, and the Republicans supported Bush. Bush's allies generally won the arguments in Congress but President Truman then exercised his veto. With US government science policy thus suspended in limbo, the OSRD was closed and its surviving programmes hived off to Defense and other government departments. Medical research, for example, went to the Public Health Service as the National Institutes of Health.

To be fair to Kilgore and Truman, there were genuine problems over Bush's plans. It seemed unprecedented for federal funds to be handed to an organization answerable only to its beneficiaries, the scientists themselves. In defending his 1947 veto Truman wrote that the bill:

would, in effect, vest the determination of vital national policies, the expenditure of large public funds, and the administration of important governmental functions in a group of individuals who would be essentially private citizens. The proposed National Science Foundation would be divorced from control by the people to an extent that implies a distinct lack of faith in democratic processes.[8]

These arguments might have gone on for ever, but, as usual, war (the Cold War this time) came to the rescue, and the National Science Foundation was created, largely along Bush's lines, in 1950, in the same year and for the same reasons as the National Security Council.

Kilgore conceded to Bush because of the lessons of the Second World War. One feature of that war had been that Congress's allocations of money had always been more than the scientists could use. There was a simple reason for this: America had run out of scientists. The military's vast demand for new technology could not be met by the researchers, whose numbers had been geared to peacetime needs. In 1940 the US's total R&D budget had been $350 million, but by 1945 it exceeded $1.6 billion, and the research personnel simply could not be expanded manyfold over five years. No one understood this better than Senator Kilgore, the chairman of the Senate's Subcommittee on War Mobilization, who in 1942, 1943 and 1945 had conducted a series of hearings on America's wartime research personnel needs.

For Kilgore, the major purpose of the National Science Foundation had never been the generation of new knowledge but, rather, the generation of trained scientists to create a strategic reserve. Consequently, in the face of the Soviet threat, he dropped his objections to Bush's model rather than forgo a foundation. The National Science Foundation's budgets were initially small, at $3.5 million annually, but they represented America's crossing of a research Rubicon – the federal government had recognized pure science as a legitimate responsibility.

SPUTNIK

Federal science budgets rose steadily during the 1950s, and then on 4 October 1957 the USSR launched a sputnik, the world's first orbiting space craft. Suddenly the Cold War was hotting up: would the Soviets destroy America from space? In the words of Wernher von Braun, a leader of the US's space programme:

> Sputnik triggered a period of self-appraisal rarely equalled in modern times. Overnight, it became popular to question the bulwarks of our society; our public educational system, our industrial strength, international policy, defense strategy and forces, the capability of our science and technology. Even the moral fiber of our people came under searching examination.[6]

Coming from von Braun, the ex-Nazi whose own moral fibre would not withstand much examination, those words are rich, but they did reflect the mood of the times. Consequently, Congress passed two momentous Acts in 1958. The National Aeronautics and Space Administration (NASA) was created, and the National Defense Education Act was passed to pour money into American education at all levels. And Washington's support for science soared. In 1940 the federal government was spending some $22 million dollars annually on pure science (corrected, for ease of comparison, to 1950 prices) but, as Figure 12.1 shows, by the late 1960s it was spending 100 times more at nearly $3 billion. Meanwhile, the federal government's total expenditure on all R&D, including defence, space, energy and medicine, had increased 500-fold by 1965, to reach $10 billion in the prices of the day.

But as Figure 12.1 also shows, the government's money then plateaued. Why? In part because certain programmes had matured

Figure 12.1 Federal expenditure on basic science and GDP per capita

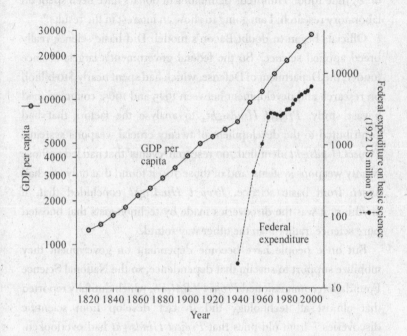

(NASA reached the moon in 1969, so its rate of budget increase declined), but partly because of government disillusion. The federal government had assumed that its support for basic science would stimulate economic growth, but where was the stimulus? Between 1820 and 1940, with only trivial government support for research, the American economy had grown at a linear rate of about 2 per cent GDP per capita per year. After 1940 the government's funding for science exploded, and the American economy grew at a linear rate of about . . . 2 per cent GDP per capita per year. The post-war economy had recovered from the Depression, of course, but that was attributable to the wartime stimulation of demand and to the peace-time resumption of international trade, not to science. As President Johnson threatened on 15 June 1966: 'Hundreds of millions of dollars have been spent on laboratory research. I am going to show an interest in the results.'

Officials began to doubt Bacon's model. Did basic science really breed applied science? So the federal government's largest science funder, the Department of Defense, which had spent nearly $10 billion on research and development between 1945 and 1965, commissioned a vast study, *Project Hindsight*, to analyse the factors that had contributed to the development of twenty crucial weapons systems. *Project Hindsight* identified 700 research 'events' that had led to those twenty weapons systems, and of those 700 it found that only two had arisen from basic science. *Project Hindsight* concluded that, if anything, it was the discoveries made by technologists that boosted pure science, rather than the other way round.

But once people have become dependent on government they mobilize support to sustain that dependence, so the National Science Foundation commissioned *Project TRACES*, which dutifully reported that almost all technology did in fact develop from scientific discoveries – from old ones that *Project Hindsight* had overlooked. *Project TRACES* thus defeated its own case for government money by going back fifty years to when basic research was privately funded but, hey, what's a little history between friends?

POPULAR SUPPORT FOR GOVERNMENT-FUNDED SCIENCE

Just as the American government was losing its faith in science, the American people adopted it – 'if we can put a man on the moon why can't we . . .'. The leader of this popular movement was Mary Lasker (now best known for the Lasker Prizes her Foundation awards; they adumbrate the Nobels). Mary Lasker had lost her husband to cancer and, though a major donor to the American Cancer Society, she decided that only with federal money would the disease be conquered, so she launched the Citizens Committee for the Conquest of Cancer, and on 9 December 1969 the Committee ran a full-page advertisement in the *New York Times* declaring 'Mr Nixon: you can cure cancer'. Mr Nixon, flattered by this confirmation of his God-like powers, naturally agreed, declaring war on cancer and pouring money into the National Institutes of Health. In his State of the Union message of 22 January 1971 he said:

> The time has come in America when the same kind of concentrated effort that split the atom and took man to the moon should be turned towards conquering this dread disease.

The legislature had already enlisted in the war on cancer, and earlier, in 1970, both the Senate and the House of Representatives had unanimously passed a resolution to cure cancer by the Bicentennial – that is, just six years later in 1976. Oddly, though, people still seem to get cancer in America.

But just as American researchers started to worry that people might detect the failure of federally funded science to meet its medical promises, OPEC raised the price of oil. President Carter responded by declaring the nation's energy problems to be the 'moral equivalent of war', and we know what governments do when they are faced with war: they raise their science budgets. In his State of the Union message for 1977 Carter said:

> Scientific research and development is an investment in the nation's future, essential for all fields, from health, agriculture, and environment

to energy, space and defense. We are pushing back the frontiers in basic research for energy, defense, and other critical national needs.

Thus does every successive president find quasi-militaristic reasons to increase his science budgets. Ronald Reagan embraced Star Wars, and though President George H. Bush cut that as part of the peace dividend, he compensated by increasing the federal support for pure science. President Clinton took office gripped by an idea called endogenous growth theory, which argued that, for economic reasons, governments should fund science – which his did, generously. George W. Bush came to office committed to restoring the old Star Wars budgets while continuing to boost the federal funding of pure science. And after 9/11 he created a new federal agency, the Department of Homeland Security, with an expanding (bio)terrorist research budget. Today, the federal government spends well over $100 billion annually on research: 50 per cent on defence, 25 per cent on health, 7 per cent on space, 6 per cent on general science, with the remainder distributed between the environment and other functions.[10.]

$100 billion is a lot of money (the 2005 figure, released in 2007, was actually $137 billion), and the struggle to control it is perennial. In his 2003 book *The Myth of Homeland Security* Marcus Ranum showed how the Homeland Security Act just happened to direct $500 million to research in universities in George W. Bush's own state of Texas. Pork-barreling motivates too much of the federal research budget: in his 1999 book *Funding Science in America: Congress, Universities and the Politics of the Academic Pork Barrel* James Savage showed how politicians are increasingly bypassing peer review to direct federal funds to their local institutions. Thus Robert Byrd of West Virginia once directed $40 million to Wheeling Jesuit College (annual budget $14 million) for a 'classroom of the future', while Senator Ted Stevens sent $40 million to the University of Alaska to extract energy from the aurora borealis (thus imitating Jonathan Swift's satire on the Royal Society in *Gulliver's Travels*, where a researcher 'had been eight years upon a project for extracting sunbeams out of cucumbers'). Meanwhile, the 2005 Transport Bill contained over 6,300 'earmarks', allocating $348 million for research and facilities at forty-two universities that were represented by politicians who were powerful enough to bypass pesky peer review.

Half a century after Vannevar Bush and Senator Kilgore first clashed over intellectual property, the politicians still clash over it. Before 1980 federal institutions kept the IPR of the science they funded, but they did little with it. They published their findings as quanta of public knowledge, but they did not commercialize them. It was the 1980 Bayh–Dole Act that transferred the IPR to the university researchers themselves who, thus incentivized, have proved to be entrepreneurs, generating over 280,000 federally funded patents and founding 2,200 firms, thus creating 260,000 jobs and adding $40 billion annually to the economy while, also, supplying the universities with licence fee income (Columbia, for example, earned $141 million in 2001). Federal labs, moreover, now collaborate with companies, which leads those companies to spend more on R&D to exploit the collaborations.[11]

Cheer-leaders for the government-funding of science claim that the private expansion of R&D budgets since Bayh–Dole has justified the taxpayers' investment, but once the cost to the taxpayer ($37 billion annually for the federal labs and $22 billion for academic science, though the two budgets overlap) is included in the appraisal, the economy enjoys no overall stimulus. Yet the success of the Bayh–Dole Act provides one more nail in the coffin of Bacon's model. Wealth is not created when quanta of public knowledge are added to earlier quanta by publication in journals in the hope that the science will be commercialized by others. Wealth is created when scientists commercialize their own discoveries. Progress is tacit, not disembodied.

* * *

The federal government's involvement in science was always ahistorical because America's unique contribution to capitalism had been its smashing of the claims of science. America invented management.

During the nineteenth century European economies were obsessed by science, technology and training, and by the end of the nineteenth century no fewer than 60 per cent of Britain's engineering workers were classified as skilled.[12] The figures in Germany or France may have been even higher, but in America they would have been much lower. Similarly, Europe was obsessed with research, and the Europeans (even the laissez-faire Britons, see later) poured money into research. But the Americans did not. Consider Andrew Carnegie's United Steel: Carnegie may have created his company by copying UK technology but he did not grow it by science, he grew it by accountancy:

Cost sheets were Carnegie's primary instrument of control. Costs were Carnegie's obsession. One of his favourite dicta was: 'Watch the costs and the profits will take care of themselves.'

A.D. Chandler, *The Visible Hand: The Managerial Revolution in American Business*, Harvard University Press, 1977

America's circumstances were different from Europe's. Land and natural resources were scarce in Europe, but they were abundant in America. Skilled labour was abundant in Europe, but scarce in America. And European markets and consumer tastes were fragmented, which demanded bespoke technologists, but the US was more homogenous. So, guided by the market, America specialized in mass production, not craftsmanship, and American industrialists developed management, not science, as their source of profits.

How the Europeans sniggered when the world's first business school, the Wharton in Pennsylvania, was founded in 1881! And how the creation of Harvard's Business School in 1908 demolished, in European eyes, its pretensions as a university! But the great American technological tome of the day was no scientific dissertation, it was Frederick Taylor's 1911 *The Principles of Scientific Management*, in which Taylor explained how the centralized direction of workers would maximize their productivity. The world's first production line may have been British (an 1804 biscuit factory in Deptford), but mass production on a line tooled by standard machinery and by interchangeable components was honed in America.

And who won the productivity race, the research-driven Europeans or the managerial Americans? S.N. Broadberry's 1997 book *The Productivity Race* shows that the American managers won. During the first half of the twentieth century America's productivity was double Europe's. So any 'explanation' of Germany's or France's overhauling of Britain economically that invokes those governments' funding of research or of education will first have to explain how the first country to overhaul laissez-faire Britain was laissez-faire America – with a productivity double that of the continent's.

In conclusion, therefore, the government funding of science in America cannot be credited with any economic or technical benefit, but it was never designed to deliver those. It happened contingently, to meet the needs of war; and only *post hoc* did people scrabble around to invent peacetime justifications for it.

Table 12.1
The Funding of Research and Development in the USA in 1940
(millions of $)

Federal funding

Agriculture	29.1
Defence	26.4
Health	2.8
Aeronautics	2.2
Others	13.6
TOTAL	74.1

States' funding

(largely agricultural)	7.0
TOTAL GOVERNMENT	81.1

Private funding

Industrially funded research and development	234
University- or foundation-funded research	31
TOTAL PRIVATE	265

Total USA funding for research and development 346.1

Sources: *Federal Funds for Research, Development and other Scientific Activities* (National Science Foundation, Washington, D.C., 1972). Industrial funding figures came from John R. Steelman, *Science and Public Policy: A Program for the Nation* (USGPO, 1947). University figures (which are for 1935–6) came from *Research – A National Resource* (USGPO, Washington, D.C., 1938).

13

Nationalizing British Knowledge

'Business is never so healthy as when, like a chicken, it must do a
certain amount of scratching around for what it gets.'

Henry Ford, *My Life and Work*, 1922

The evolution of British government science policy was similar to
America's. After 1688 Britain was laissez-faire in science, leading
the world through the Agricultural and Industrial Revolutions in the
absence of government money for science until war pushed the
government into funding science – to no economic benefit. Whitehall's
bitter experiences with Babbage long deterred politicians from further
involvement with research, and though Whitehall did accept
responsibilities for certain functions that involved research, that
research was always mission orientated. A celebrated example was the
chronometer.

During the early eighteenth century, as trade grew, so did the
numbers of shipwrecks, so in 1714 the government established the
Board of Longitude to reward with £20,000 the inventor of a method
of determining it. By 1735 John Harrison, a watchmaker, had
developed an excellent chronometer, and in 1762 a ship sailed to
Jamaica carrying a Harrison chronometer that registered only a 5
second error for the voyage, thus allowing longitude to be determined.
Yet, in a classic example of Shakespeare's 'insolence of office', the
government delayed payment until 1773, and then only because
Harrison went to court. The government funding of science often
disappoints.

Britain being a world power, the government supported science in support of its mercantile world view. So the Ordnance Survey, which still makes the maps we all use in Britain, was created in 1745 as a defence initiative (hence the military name) after the Duke of Cumberland struggled, following the Battle of Culloden, to root out the Jacobites hidden in the unmapped glens. And the Royal Navy supported imperial horticulture: Captain Bligh of HMS *Bounty* was transporting breadfruit plants from Tahiti to the Caribbean to feed the slaves when Fletcher Christian cast him off. The Royal Navy also supported imperial cartography: HMS *Beagle* was mapping the southern ocean shores when it offered Mr Darwin a berth.

But those research activities cost trivial sums of money. In 1849 Parliament did, admittedly, vote an annual grant-in-aid of £1,000 to the Royal Society, but that was just a token of esteem. Gradually, however, the pressure for greater support from government began to build up, the decisive event being the 1867 Paris Exhibition, which, as we saw earlier, was widely – if wrongly – believed to have exposed British scientific weakness.

By 1867 the government was still ignoring science, and the civil estimates showed that all government departments together spent less than £400,000 a year on research, the money going only on museums, gardens, observatories and other scientific work that directly supported government functions. No pure science was supported. But Charles Babbage had created the British Association for the Advancement of Science (generally known as the British Association) in 1831 to campaign for government money, and the British Association fuelled the scare stories of the 1867 Paris Exhibition. The most popular talk of its 1868 meeting was 'On the Necessity for State Intervention to Secure the Progress of Physical Science' by Colonel Strange, an amateur astronomer.

Other scientists kept up the pressure. The most powerful advocate was Norman Lockyer (1836–1920). A good astronomer (he discovered the element helium), Lockyer was employed by the government's Science and Art Department. In 1869 he founded *Nature*, and though *Nature* has always published scientific papers, Lockyer forged it primarily as a political platform. His favoured tool was to proclaim the collapse of British science. In 1872, for example, he announced: 'England, so far as the advancement of knowledge goes, is but a third-rate or fourth-rate power.' Tragically, things got even worse, and in 1873 he claimed that 'science is all but dead in England', which must have come as a shock to Clark Maxwell and William Kelvin, then the world's leading physicists, and to Darwin and Huxley, then the world's leading biologists. (*Nature* still publishes histrionics, and on 22 March 1990 it proclaimed that the British government had 'all but killed off' British research, Britain then being second only to the USA in the numbers of scientific papers it published and in the numbers of citations those papers attracted.)

Not all scientists joined the campaign, and no less a giant than Alfred Russel Wallace (1823–1913), the co-discoverer with Darwin of evolution through natural selection, observed that the free market or 'public competition' was providing more and better jobs in science than could government:

> Experience shows that public competition ensures a greater supply of the materials and a greater demand for the products of science and art, and is thus a greater stimulus to true and healthy progress than any government patronage.
>
> Peter Raby, *Alfred Russel Wallace*, Chatto and Windus 2001.

Wallace, who had earned his living by collecting plants and animals from the tropics for collectors in Britain, did, however, eventually accept 'government patronage' in the shape of an old age pension from

Parliament in recognition of his fame. (Those pensions could be troublesome. When Parliament awarded a pension to Edward Jenner, the discoverer of vaccination, it was denounced by Mrs Jenner no less, who argued that her husband could have earned a good living had he deigned to vaccinate patients himself instead of frittering away his time on research.)

While Gladstone was prime minister, the lobbying for science was ignored. Gladstone accepted that science might be a public good, but, like Rumford, he believed that it was for rich individuals, not the state, to provide public goods. Consequently, William Cavendish, the seventh Duke of Devonshire, rose to Gladstone's challenge, endowing the Cavendish Laboratories in Cambridge in 1871 with no less than £8,950. The Cavendish has since proved to be one of the world's finest laboratories, having supported such giants as J.J. Thomson, who discovered the electron, Ernest Rutherford, who split the atom, and James Watson and Francis Crick, who discovered the double helical structure of DNA – a tribute to the private endowment of science.

But in 1874 Gladstone lost the general election, and the incoming Tories thought in Baconian terms of superpower rivalry. Moreover, under Disraeli's leadership, they subscribed to corporate beliefs similar to Congressman Morill's; they opposed the repeal of the corn laws but they supported subsidies. So in 1876 the Tories voted £4,000 a year for the Science and Art Department to fund pure science. But between 1876 and 1913 that sum did not rise. What did rise, however, was an extraordinary backlash, as public opinion, led by Christians and humanitarians, turned against science. Christians had been appalled at the damage that the theory of evolution had perpetrated against their faith, and by 1904 the Christian backlash against science was so fierce that the government's Education Department, in its 'Regulations for Secondary Schools', abolished two-thirds of the curriculum in science and forbade any further teaching of technology in government schools. Humanitarians, meanwhile, were appalled by the findings of the Royal Commission on Vivisection of 1876, which revealed that British university scientists treated animals with casual cruelty.

Vivisection has a long and shameless history. The first great book in modern medical science, Vesalius's *Fabrica*, the anatomical text published in 1543, contains on page 662 a drawing of a living unanaesthetized pig, stretched out by chains and screaming in agony as it is dissected. The Royal Society, too, performed vivisection, leading Robert Hooke to complain in November 1664 that 'I shall hardly be induced to make further trials of this kind, because of the torture of the creature', but the Fellows persisted.

The heartlessness of scientists was, of course, one of the drivers of the Romantic Movement, and in his 1750 European bestseller *Discours sur les arts et les sciences* Jean-Jacques Rousseau had denounced science as the tool of greedy tyrants, and he had portrayed scientists as unfeeling brutes. By 1818 his disciple Mary Shelley would write *Frankenstein* to portray scientists as vainglorious, emotional cripples. Meanwhile, Samuel Taylor Coleridge stated that 'poetry is opposed to science', William Blake maintained that 'Art is the Tree of Life . . . Science is the Tree of Death', and in his *Tables Turned* William Wordsworth wrote:

> Sweet is the love which nature brings;
> Our meddling intellect
> Mishapes the beauteous form of things:
> We murder to dissect.

Faced with such sentiments and with the revelations of the 1876 Royal Commission on Vivisection, British legislators resisted giving more money for research. In any event, many scientists themselves repudiated the British Association's campaign for state funding, and in 1880 they created the Society for Opposing the Endowment of Research (secretary William Noble FRS) to block 'the wretched whining cry for State funds in aid of research'. On 25 February 1881 Sir George Airy, the Astronomer Royal, outlined the Society's policy:

> Successful researches have in nearly every instance originated with private persons, or with persons whose positions were so nearly private . . . that the investigators acted under private influence, without the dangers attending conjunction with the State.
>
> *English Mechanic*, 25 February 1881

And revulsion was felt at the behaviour of Lockyer who, though the employee of the Science and Art Department, had led the campaign for government money by his editorship of *Nature*. In 1876 the government voted the Science and Art Department its £4,000, and what did the Department do with it? It created a full-time research post and a new solar physics laboratory for . . . Norman Lockyer! Just as in America, it was to be the public advocates of government funding who were to most benefit from it.

Or was it? The croneyism shocked everybody, and the public outrage was so fierce that the government took fright, retrieving its £4,000 and transferring it as a further grant-in-aid to the Royal Society, to be distributed by open competition by a grant committee. But many distinguished scientists refused to join the committee. William Flower, the director of the Natural History Museum, refused: 'The large increase of this method of subsidizing science, accompanied as it is with the (as it appears to me) humiliating necessity of personal application in each case, must do much to lower the dignity of recipients and detract from the independent position which scientific men ought to occupy in this country.' The distaste (and the suspicion that the members of the committee simply funded each others' research) was so pervasive that when, in 1881, the Education Minister, A.J. Mundella, offered the Royal Society a further £2,000 a year to fund research fellowships, the Society refused it with a collective shudder, as adumbrated by Richard Proctor in his 1876 book *Wages and Wants of Science Workers*: 'Few circumstances have caused true lovers of science in recent times more pain than the outcry for the Endowment of Research.'

Meanwhile, the private endowments for science continued to pour in. When Alfred Mond endowed the Davy–Faraday Laboratory at the Royal Institution, the *Daily Telegraph* noted on 4 July 1894 that the Mond bequest had done 'more for physical science at one stroke than all the Cabinets of Her Majesty have done since the commencement of that reign, one of the greatest glories of which has been the advance of research in England'. When the Imperial Cancer Research Fund opened in 1901, and the Jenner (now Lister) Institute for Bacteriology soon after, Britain's privately funded biological research soon matched its physical. And the economy continued to grow.

As ever, it was war that prompted the government funding of science. The tipping point was the Boer War (1899–1902), during which the army had rejected 40 per cent of all potential recruits as unfit. That figure shocked the nation, which had assumed that almost all young Britons would be fit enough to fight. But people were then falling into the grip of eugenics, and the figure of 40 per cent only confirmed that the superior white race was degenerating, even at the heart of the British Empire.

EUGENICS

Eugenics represents one of modern science's most horrific perversions, and we all try to forget it now or blame it on Hitler, but eugenics dominated the science policies of the first half of the twentieth century. One reflection of the horror eugenics still induces is that its most readable history, *Intellectuals and the Masses* (1992), was written not by a scientist but by a professor of literature, John Carey of Oxford University. To show how pervasive eugenic ideas once were, let me extract from Carey's book some quotes from well-known authors:

I believe that the mob, the mass, the herd, will always be despicable.

Gustave Flaubert, 1871

The swarms of black and brown and dirty-white and yellow people have to go. It is their portion to die out and disappear.

H.G. Wells, 1901

Three cheers for the inventors of poison gas.

D.H. Lawrence, 1921

Extermination must be put on a scientific basis if it is ever to be carried out humanely and thoroughly . . . if we desire a certain type of civilization and culture, we must exterminate the sort of people who do not fit into it.

George Bernard Shaw, 1933

The eugenics movement (the term was coined by Francis Galton, a cousin of Darwin's, in 1883 from *eu gene* or well born) was born of snobbery. Gregory King had shown in the seventeenth century that the working classes were being culled by poverty and starvation, but the food and wealth of the Agricultural and Industrial Revolutions had since changed the demographics, and the proles were now breeding. As H.G. Wells complained: 'The extravagant swarm of new births was the essential disaster of the nineteenth century.' Because people understood that evolution represented the 'survival of the fittest', they feared that unfit proles were being artificially sustained. As W.B. Yeats wrote in his 1939 *On the Boiler*, quoted by Carey:

> Since about 1900 the better stocks have not been replacing their numbers, while the stupider and less healthy have been. Since improvements in agriculture and industry are threatening to supply everyone with the necessities of life and so remove the last check upon the multiplication of the ineducable masses, it will become the duty of the educated classes to seize and control one or more of these necessities. A prolonged civil war seems likely.

As the social pressures for eugenics grew, so the pressure on scientists to create a research discipline in its support grew too, and when in 1900 Mendel's laws of inheritance were rediscovered, they were believed to have 'proved' that society's problems were genetic. Thereafter, eugenics research accelerated, and in 1904 alone Charles Davenport (1866–1944), who had been successively a professor of biology at Harvard and Chicago Universities, established the Cold Spring Harbor Laboratory for eugenics research on Long Island (the money having been provided by the Carnegie Foundation), Francis Galton endowed a eugenics professorship at University College London, and the German psychiatrist Alfred Ploetz, who in 1895 had published *The Fitness of our Race* ('destroying unworthy life is purely a healing treatment'), established the research Society for Racial Hygiene.

For as long as eugenics research remained in the laboratory, it could be useful. Davenport determined that eye and hair colour were inherited in a Mendelian fashion, while in Germany Otmar von Verschuer, the director of the Kaiser Wilhelm Institute of Anthropology, Human Heredity and Eugenics (which had been founded in 1927 with money provided in part by the Rockefeller and Loeb Foundations from the US), studied the inheritance of susceptibility to infectious diseases. Ever since the Native Americans had succumbed in their millions to European diseases, such as measles, scientists had understood that different races had inherited different susceptibilities to different diseases, and by the careful analysis of the incidence of TB between identical and non-identical twins in Germany, Verschuer chronicled the genetics. One of Verschuer's pupils was Josef Mengele, whose early research on the inheritance of cleft lip in Asia and Europe was creditable.

But when governments adopted eugenics, public squalor followed. In his 1912 book *Heredity and Eugenics* Charles Davenport had written that 'the feebleminded, drunkards, paupers, sex offenders, and criminals are the victims of a recessive gene', and the argument persuaded many politicians that, society's problems being genetic, they were soluble by sterilization. By 1935 no fewer than twenty-seven states in the US had passed compulsory sterilization laws; indeed, by 1935 California alone had sterilized 9,931 mental defectives. By 1939 over 100,000 citizens across the US had been compulsorily sterilized. During the 1930s, moreover, British Columbia in Canada, all the Scandinavian countries, the Baltic countries, Germany and Switzerland had passed laws to allow the compulsory sterilization of mental defectives and other deviants, including single mothers.

Practice differed from country to country, but the German Racial Hygiene Courts were typical. An individual was arraigned for being stupid or anti-social before three judges, two of whom were doctors, one of whom was a lawyer, and if found wanting the individual was compulsorily sterilized.

Among the Anglo-Saxon countries, only the Netherlands and Britain resisted compulsory sterilization on the grounds of civil liberties (Churchill, who had 'wanted the curse of madness to die', was disappointed when the Commons rejected the bill). It was only the Catholic countries that systematically opposed eugenics, for they opposed any scientific intervention into fertility. As the Catholic convert G.K. Chesterton was to write: 'Eugenicists have discovered how to combine hardening of the heart with softening of the head.'

But the scientists had hard hearts because their purses were engaged. Consider the comments of Herbert Spencer Jennings, a professor at Johns Hopkins University. Jennings was a good biologist (he had studied spinal nerve reflexes) and was troubled by eugenics, which he criticized within the professional journals, but in his 1930 book *The Biological Basis of Human Nature* he rejoiced that it had put biology on the political map:

> Gone are the days when the biologist used to be pictured as an abused creature, his pocket bulging with snakes and newts. The conduct of society is to be based on sound biological maxims! Biology has become popular!

Scientists rarely criticized eugenics because it was their source of grants and of influence: Jonathan Harwood, the historian of science, could find not one German scientist who publicly criticized the eugenics movement.[1] In the English-speaking world Lionel Penrose, who actually held the chair of eugenics at University College London, did publicly criticize eugenics, but only for its unethical application, not the discipline itself.

Yet the underlying premise behind eugenics, namely that the sterilization of unwanted individuals would clear society of their recessive genes, was soon disproved. As early as 1917 the Hardy-

Weinberg Principle, a mathematical analysis of the inheritance of recessive genes, had established that mass sterilization simply could not clear society of those genes. The genes are called 'recessive' because they are mostly carried in a hidden form by large numbers of normal people, not by the relatively small number of diseased people. So, for example, only one in 2,000 Europeans is born with the disease of cystic fibrosis, but no fewer than one in twenty normal people carry a copy of the gene. The sterilization of the rare diseased individuals, therefore, will not prevent the large numbers of normal people from reproducing the genes (especially as almost all patients with cystic fibrosis are sterile anyway, which has nonetheless not stopped the disease from propagating). Yet the disproof of their science did not inhibit the eugenicists.

HOW SCIENTISTS WORK

In the face of the disproof of Hardy–Weinberg, most scientists still believed in eugenics; today, in the face of the OECD's and others disproofs, most scientists still believe that research needs to be funded by the state. Let me here, therefore, explore how and why scientists ignore disproof.

Non-scientists believe that science works according to Bacon's ideas of induction. Non-scientists believe that data is collected dispassionately and that a theory is then proposed to explain it. But, as Einstein pointed out, 'it is theory which decides what we observe'. Scientists do not collect data dispassionately – there is so much raw data out there, where would a person start? Does a scientist start by correlating what Namibians eat for breakfast with the spin of subatomic particles? No, a scientist starts by building on the collective judgements of other scientists, just as – when witnesses are asked to describe a traffic accident – they don't start by describing everything they did or did not see, instead they describe what the cars were doing. Scientists, like the rest of us therefore, start with a theory of what is important, which determines the data they collect.

It is because scientists must select their observations that, inevitably, they tend to reinforce their own preconceptions. In the process that Thomas Kuhn called 'normal science' in his 1962 book *The Structure of Scientific Revolutions*, scientists generally select their observations to the established paradigm. So, for example, medieval astronomers fitted their findings to the paradigm that the earth was the centre of the universe, which was why they described the orbits of the planets as circles-within-circles or epicycles.

But, in a famous episode, Nicolaus Copernicus dissented. Copernicus (1473–1543), a Catholic priest, argued that God loved purity – and what could be purer than a circle and what could be less pure than an epicycle? So in 1543 Copernicus selected his own astronomical observations, to show that the sun was at the centre of the universe. The key point was that Copernicus had no evidence to disprove epicycles, indeed the best evidence then available contradicted Copernicus, and the leading astronomer even of the subsequent generation, Tycho Brahe (1546–1601), continued to generate planetary observations that contradicted Copernicus and supported epicycles.

Non-scientists believe that science works according to Karl Popper's ideas of falsifiability. Non-scientists also believe that, when contradictory data is collected, the old theory is discarded for a new one. But the use of the same data by the followers of Copernicus and Brahe to support two incompatible paradigms reveals how scientists really work – by verification not by falsification. Scientists create hypotheses based on hunches, sometimes unsavoury ones – 'God loves purity', 'Black people are inferior to white people', 'DNA, which is a sugar, carries the genetic information' – and then they seek to prove them.

In the past, paradigm shifting could be horrible as competing protagonists railed against each other, but scientists have become better at paradigm shifting, and contemporary revolutions, such as DNA or the internet, will now arise by simple discovery. Nonetheless, Einstein's dictum that theory determines what we observe still describes science, and it still explains eugenics. During the nineteenth century the European nations acquired vast empires, the oppression of whose natives required justification. The Europeans' mutual competition, moreover, provoked anxieties over the quality of the domestic cannon fodder. And immigration was a worry. Irving Fisher, the great Yale economist, wrote to Charles Davenport that: 'The stresses of immigration alone provide a golden opportunity to get people in general to talk eugenics.'[2] The eugenicists shared those anxieties, which they alleviated in concepts of racial superiority, which they then sought to verify scientifically. So when they encountered the Hardy–Weinberg Principle, they did not cry 'falsifier, pants on fire!', they simply ignored it.

Scientists, including the best ones, are always ignoring inconvenient data. Because they are working at the limits of knowledge, they have to. Consider the earlier controversy over the age of the earth. In the nineteenth century geologist Sir Charles Lyell (1797–1875) had, by his study of the rate of erosion of cliffs, proposed the earth to be hundreds of millions of years old. But, as we know from volcanoes, the core of the earth is red hot. And when geologists measured the temperature of the molten core, and when they calculated its rate of heat loss, they concluded that the earth could be only a few millions of years old. Had it been any older, its core would have completely cooled. Lyell had apparently been falsified.

In the face of this apparent falsification, did Lyell's followers ditch their ideas? No. They knew how old the sedimentary rocks had to be, and they didn't believe the falsifiers, but not knowing how to falsify the falsifiers, they simply ignored them, assuming, Micawber-like, that something would turn up. Which it did. Somebody in some other discipline discovered radioactivity, somebody discovered the core of the earth to be radioactive, somebody discovered that radioactive reactions emitted heat, and, hey presto! the problem was resolved: the core of the earth generates heat, which is why it is still hot, and the earth is indeed very old.

So the eugenicists trusted: the basis of their science had apparently been falsified, but they still faced the problem of mental defectives or of single mothers, they still faced the yellow peril and the black natives, and they still faced the potentially hostile German/French/British army (delete as appropriate), so why should they not sterilize their unwanted domestic detritus while waiting for the new science that would hopefully emerge to falsify Hardy–Weinberg's apparent falsification?

It is because, ultimately, science is subjective, driven by hunch and prejudice and selected data, that the postmodernists claim that science is only a 'social construct' and that there can be no 'objective truth'. And because science is controlled by the funders of the grants and the publishers of the journals, so they claim that science is only what a handful of powerful cronies decide it is. In the words of the Austrian-born philosopher Paul Feyerabend:

> The scientist is restricted by his instruments, money, the attitudes of his colleagues, his playmates, and by innumerable physiological, sociological, historical constraints.
>
> *Against Method*, 1975

Much evidence supports the postmodernist argument that science is political. Consider the career of Bjorn Lomborg from Denmark who, in his 2001 book *The Skeptical Environmentalist*, inverted a conventional paradigm. Environmentalists generally teach that economic development destroys the environment, but Lomborg proposed the opposite: the richer the country, the better it conserves its environment.

Lomborg was inundated with abuse (*Scientific American* dedicated 11 pages of its January 2002 issue to 'Science defends itself against *The Skeptical Environmentalist*', though, ironically, the same issue carried an article on a new generation of power stations that would be pollution-free, thus showing how the new technologies of rich countries are indeed environmentally friendly), and he left the University of Aarhus, whose researchers had maintained a website of anti-Lomborg denigration. Meanwhile Denmark's Committee on Scientific Dishonesty condemned him for 'dishonesty' and for flouting 'good scientific practice'.

But Lomborg was then provided with his own environmental institute by the Danish government (headed by Prime Minister Rasmussen, a neo-con), which dismayed those environmentalists who had denounced Lomborg to protect their own governmental grants. And in 2004 the Danish Ministry of Science, having assessed the evidence, concluded that Lomborg's book was good science and that the Danish Committee on Scientific Dishonesty had been wrong to condemn it. Indeed, the Ministry in its turn condemned the DCSD for 'unsatisfactory and emotional conduct deserving of criticism'.

But, while the postmodernists are right in viewing science as political, they are wrong to be ultimately pessimistic about scientific truth. As long as there is a multiplicity of funders, there will always be little boys like Lomborg who, as honest reporters, will eventually show up the big men of science as having no clothes. Consequently, competing paradigms will continue to be generated.

Yet not even Lomborg can be trusted. No scientist, as an advocate, can be. There is a class of drugs known as calcium-channel antagonists, which are used to treat heart disease. Over the years at least seventy clinical studies on these drugs have been published by university professors and practising doctors. Some of those studies were funded by the manufacturers of those drugs, while others were funded by independent sources, including charities, government research agencies and hospitals. In 1998 a group of investigators from the University of Toronto studied those seventy different papers, discovering:

> A strong association between scientists' opinions about safety and their financial relationships with the manufacturers. Supportive scientists were much more likely than critical scientists to have financial associations with the manufacturers.
>
> H. Stelfox *et al., New England Journal of Medicine*, 1998, 338:101–6

University professors and practising doctors, therefore, publish findings that support their sources of money. Repeated surveys of scientists' publications have confirmed this finding, and though the medical journals now insist that authors publicly acknowledge their funding sources, such acknowledgements seem not to affect the scientists' conclusions. 'Twas ever thus, and as early as 1667 Thomas Sprat noted in his *History of the Royal Society of London*: 'For whosoever has fix'd on his Cause, before he has experimented; can hardly avoid fitting his Experiment, and his Observations, to his own Cause, which he had before imagin'd; rather than the cause of the truth of the Experiment it self.'

Thus we see here the paradox that science, which appears to be the most honest of human activities, will be promoted in a partisan fashion. Science is unpredictable – nobody knows how any particular project will develop – so no funder can know which, of several competing projects, to support. Advocacy is therefore an integral part of science, and in practice the projects that get funded are those whose advocates are the most effective – or shameless. And verification – the deliberate ignoring of inconvenient data – is an integral part of science.

Which is why the once-near-universal triumph of eugenics says something important about 'science'. 'Science' is not a collection of dispassionate public facts *à la* Bacon. 'Science' is what the scientists say it is, and if almost every prominent scientist believes in eugenics – and believes it so passionately that the belief will spillover to non-scientists such as H.G. Wells, D.H. Lawrence and George Bernard Shaw – then that is what 'science' is. There is no such thing as 'science', there are only scientists.

We see here, therefore, a huge destruction of Bacon's ideas of science. Bacon believed that science was dispassionate, with scientists collecting a mass of objective data dispassionately and inducing theories dispassionately. Bacon saw induction as almost mechanical in its workings. Under those circumstances its funding by government might be helpful. But if science is, in fact, factional and verificational, then its funding by government only aggravates those problems because government must inevitably favour one faction over another. And because politicians (even in democratic countries) are effectively unaccountable in many of their funding decisions, their funding removes science from those ultimate tests of credibility, namely the collective judgements of the market, civil society and of the disinterested parts of the scientific community.

And in 1913 the British government, worried about war looming in Europe and with 40 per cent of the country's youths apparently unfit, created the Medical Research Council. No responsible government could ignore the threat to Britain's racial hygiene.

THE MEDICAL RESEARCH COUNCIL

It was Lloyd George, then Chancellor of the Exchequer, who established the MRC in 1913, with a budget of £56,000 a year, taken from the mandatory weekly contributions that working men were then making to their medical insurance. Lloyd George instructed the MRC to prioritize research into tuberculosis. The historians of the MRC have wondered why, because by 1913 much research was already being funded by three large voluntary organizations, the National Tuberculosis Association (USA), the National Association for the Prevention of Tuberculosis (UK) and the Tuberculosis Society, later Association (UK).[3] Moreover, the Germans also were active, Robert Koch having discovered the tubercle bacillus in 1882. But David Lloyd George's father had died of TB and, fearing he might have inherited a susceptibility, he identified it as the Nation's Number One Medical Research Priority. David Lloyd George was a politician who treated the organs of the state (and the women contained within) as his chattels.

Yet Lloyd George moved on, as did the MRC, which soon ignored tuberculosis. Instead, it settled into becoming the instrument of Walter Morley Fletcher, its secretary from 1914 until his death in 1933. An uninspired physiologist from Cambridge, Fletcher had not been the obvious choice for the MRC, but the first three choices had turned it down. And Fletcher was dreadful. We have seen how the advocates of the government funding of science, men like Vannevar Bush or Norman Lockyer, had tried to become tzars of science, and we have also seen how the politics in which they had to engage forced accountability on them. No such accountability troubled Fletcher. The government had created the MRC, and he slipped in as an anonymous civil servant. Yet Fletcher suffered none of the civil servant's constraints because the organization to which the MRC reported, the Committee of the Privy Council for Medical Research, met only once during his lifetime![4] Talk about government (as opposed to market) failure.

Fletcher ruled the MRC as a tyrant, believing that he should control all medical research in the country. He loathed the autonomous medical research charities, stating in 1918 that:

> The Medical Research Committee . . . is able to organize and coordinate work at every centre of medical research in the Kingdom . . . If any fresh private endowment for medical research is available, it will be very important either that it should be administered by the Medical Research Committee, or, if it be under a separate trust or management, that this should be in close relation to the Committee's work.

Nonetheless, despite Fletcher's campaigns against them, some medical charities persisted in their independent research, and others continued to be founded. Despite his public hostility, for example, the Cancer Research Campaign (CRC) was soon founded, and by 2001 it was spending £56 million annually on research, making it the third largest funder of medical research in Britain after the MRC (£345 million) and the Wellcome Trust (£279 million). The Imperial Cancer Research Fund, with which the CRC has since fused, was also spending £56 million in 2001. Indeed, medical research in Britain is now dominated by the charitable sector – a sector Fletcher tried to crush – which spent £543 million in 2001.[5] Thank heavens he failed.

Fletcher, moreover, loathed practising doctors, so the MRC broke all contact with the Royal Colleges, the professional and training organizations for British doctors, because they were 'unfit', 'incompetent' and 'idiotic'. And Fletcher hated public health: when the Ministry of Health initiated research into public health, Fletcher forbade any member of the MRC from communicating with the Ministry. The only organization in Britain that Fletcher treated generously, pouring money at it, was the Department of Biochemistry in Cambridge, whose head was Gowland Hopkins, an old friend and vitamin expert.

Yet Cambridge biochemistry was then going through a frivolous phase, and it was dominated by men like J.B.S. Haldane (a communist who wrote more for the *Daily Worker,* which he edited, than for the *Biochemical Journal,* which he didn't) who dissipated his energies on the house journal *Brighter Biochemistry*, which was full of 'jokes' like these 'confessions' of Haldane's in the issue for 1929–30:

> *When are you at your best?* My optimum temperature is 0°C, my optimum pH is a question for L.J. Harris, being obtained by saturating a 25% aqueous solution of ethanol with CO_2. (Yes, Mr Editor, you can get iced champagne for 4/- a bottle in Paris.)
>
> *When are you at your worst?* There is no evidence that the depths of my potential iniquity have been plumbed. But probably at the end of an hour's lecture in French, when my notes have only lasted for 45 minutes.
>
> *Do you think life worth living?* Yes, but I do not think that the majority of resting bugs, dons and bacteriophages are alive. My answer only applies to higher organisms.

By 1927, frightened by what working men might think they were getting for their compulsory contributions to the MRC, Fletcher complained that:

> Having somehow bagged the credit for inventing vitamins, Hopkins spends all his time collecting gold medals on the strength of it, and yet in the past ten years has neither done, nor got others to do, a hand's turn of work on the subject. His place bristles with clever young Jews and talkative women, who are frightfully learned about protein molecules and oxidation-reduction potentials and all that. But the

277

vitamin story is clamouring for analysis . . . yet not a soul at Cambridge will look at it.[6]

Fletcher could not cut the MRC's funding for Hopkins's department, for that would have admitted error, so in 1927 he persuaded the trustees of the Dunn Estate (a private medical charity!) to endow a Dunn Nutrition Laboratory in Cambridge to do the work on the vitamins that Hopkins would not do. Yet advances in vitamins continued to elude Cambridge, vitamin B12 for example being discovered by Lester Smith of Glaxo.

But the MRC's incompetencies lay in the future; let us return to 1913, or rather to 1914–18 because, as ever, it was war that catapulted the government into upping its funding of science. In 1916 the Department of Scientific and Industrial Research, the precursor of the UK government's physical science research councils, was created with a huge budget of £1 million to improve poison gas, barbed wire and tanks, the three great needs of Flanders. As importantly in the long term, the precursor of the Higher Education Funding Councils was created in 1919 as the University Grants Committee.

THE STATE FUNDING OF BRITISH UNIVERSITIES

After 1688 the English universities were left to their own devices, and they were self-supporting. They received government money only for specific tasks (London University received £3,320 in 1841, for example, for acting as the central examination body for the colonies). The Scottish government, on the continental model, did support its universities, and after the Union of 1707 those commitments were continued, but they did not amount to huge sums (£5,077 in 1832, for instance). But laissez-faire worked, and between 1851 and 1892 eleven university colleges were founded privately in England and Wales. Typical was Mason College, later Birmingham University, endowed by Josiah Mason, a local industrialist. On laying the foundation stone in 1875 he said: 'I, who have never been blessed with children of my own, may yet, in these students, leave behind me an intelligent, earnest, industrious and truth-loving and truth-seeking progeny for generations to come.'

On being privately founded, however, the university colleges soon united to press for government money. In 1887 they mounted a joint campaign, bombarding the press with the two standard arguments – 'the other country' and 'the public good' – that activists still use. On 21 March 1887 the biologist Thomas Huxley, warning of Germany as 'the other country', wrote in *The Times*: 'The latter years of the century promise to see us in an industrial war of far more serious import than the military wars of its opening years.' And on 1 July 1887, claiming the universities to be a public good, Sir John Lubbock wrote in *The Times*: 'The claims of these colleges were not based alone on their service to learning and study; they were calculated to contribute largely to the material prospect of the country.'

To appease the universities in 1889 the Government created an ad hoc Committee on Grants to University Colleges, but its budgets were always small: by 1912 the committee was spending £150,000 annually, but that accounted for only some 10–20 per cent of the civic universities' income. Oxford and Cambridge, meanwhile, remained totally independent, despising the universities of the other country. A nineteenth century Oxford ditty, written by philosopher Henry Mansel sneered:

> Professors we, from over the sea,
> From the land where Professors in plenty be,
> And we thrive and flourish, as well as may,
> In the land that produced one Kant with a K,
> And many Cants with a C.[7]

The First World War changed everything, however, and in 1918 the universities, including even Oxford and Cambridge, sent a deputation to the Treasury, begging near-hysterically for money. Their leader, Sir Oliver Lodge, principal of Birmingham University asked 'for a doubling of the grant now and a doubling soon . . . we cannot wait; we want these two doublings put together, we want a quadrupling at once'. For once, the universities' claims were not false – they had been bankrupted by the war.

About half of the universities' income before the war had come from their investments (largely, in those non-inflationary days, in fixed-interest vehicles such as Consols) but the war had unleashed a terrible inflation, and by 1920 the after-tax purchasing power of Consols had fallen to only 26 per cent of their 1914 value. And the universities had lost their fee income too, because for four years almost all young men had gone straight to the Front. It was, therefore, genuinely to save the universities from bankruptcy that in 1919 the Universities Grants Committee was instituted with an annual budget of over £1 million. By 1921 the UGC's grant was £1,840,832. Since over half of the money went on science, this represented massive government support for research.

GERMANY AS THE OTHER COUNTRY

The deputation, on begging the government for money, had complained that 'the German universities are sustained by a State Grant averaging 72 per cent of their total income'. But that argument was dangerous because the German universities, being the instruments of the Prussian and German states, have a shameful record.

When Germany invaded neutral Belgium in August 1914, soldiers destroyed the library of the University of Louvain. That had been founded in 1425, and its destruction led to the loss of 300,000 books and over a thousand medieval and original manuscripts. The world was shocked, but Adolf von Harnack, the doyen of German theology, gathered ninety-two of the most distinguished German scholars, including Max Planck the physicist, to justify the burning of the library on racial grounds: 'The enemy are Russian hordes allied with Mongols and Negroes unleashed against the white race.'[8] In 1915 another 352 of the country's most prominent professors signed the Declaration of Intellectuals to support German war aims, which included the creation of a huge empire within Europe.

Only four German academics publicly dissented from the Declaration, one of whom, interestingly, was Albert Einstein. Einstein was no saint (he beat both his wives, he neglected his children, and he womanized), yet because his research had not been funded by the State, he felt free to criticise it. We see here a message for academic freedom: the German universities, being dependent on the State, were bound to it, which is why those German and Austrian thinkers who emerged with honour between 1914 and 1945 were generally the autonomous writers such as Thomas Mann or Stefan Zweig, not the professors.

And if the German universities' support for the Kaiser was strong, for Hitler it was passionate. In his *Rektoral* address at the University of Freiburg, the philosopher Martin Heidegger called on his colleagues to recognize the Führer as the leader whom destiny had called to save the nation. By 1933 – before they needed to – over half of all university professors in Germany had joined the Nazi Party, as had over half of all academic biologists, rendering those the professions with the greatest Nazi Party penetration in all of Germany. The academics so loved Hitler that by 1947, even after over 4,000 had been dismissed for active Nazism, more than half of the remaining academics were still ex-party members.

After the war Marburg University was only one of several universities that elected an ex-Nazi (the chairman of a Nazi military court) as *Rektor*. But the universities and Hitler agreed on eugenics, nationalism and the state funding of science and education, so they supported each other in mutual esteem.

NATIONALISING THE BRITISH UNIVERSITIES

By 1921, when local education authority grants were added to the University Grants Committee's £1.8 million, more than 50 per cent of the universities' income was coming from government bodies. The universities were effectively being nationalized, and the proportion of their income that came from government continued to rise. Though both government and the universities justified the money on the grounds that British industry needed more scientists, industry itself did not agree. The Board of Education (1909), the Balfour Committee (1929), the Malcolm Committee (1929) and the deputy under-secretary at the Board of Education (1942) had all reported that the supply of scientists and technologists had fully met, and even exceeded, industrial demand.[8] But as John Maynard Keynes said:

> The ideas of economists and political philosophers, both when they are right and when they are wrong, are more powerful than is commonly understood. Indeed the world is ruled by little else. Practical men, who believe themselves to be quite exempt from intellectual influences, are usually the slaves of some defunct economist.
>
> *The General Theory of Employment, Interest and Money*,1935

Government and the universities were then slaves to a very defunct Bacon, and the proportion of the universities' income that came from government continued to rise. In 1945 the annual UGC grant rose to £4.1 million, being £25 million by 1957. The numbers of academics kept on doubling, from 5,000 in 1938–9, to 11,000 in 1954–5, to 25,839 in 1967. Today the British universities employ some 300,000 people, nearly half of whom are academics.

John Murray, the principal of Exeter University College, lamented in 1935 that: 'A university is like a man, it may gain the whole world and lose its own soul.' Consequently, in May 1946, when the government's Barlow Committee on Scientific Manpower demanded that 'the State should increasingly concern itself with positive University policy', the Committee of Vice Chancellors and Principals, instead of rebuffing the government, replied that 'they will be glad to have a greater measure of guidance from the Government'. When, on 23 March 1987, Sir George Porter, the president of the Royal Society, wrote in the *Independent* asking for more government money for science, he also asked for greater government control:

What Britain needs is a clear and visible long-term policy for the whole of science and technology, determined and accepted at the highest level: a policy which co-ordinates both science education and research, universities and industry. Many of us in this country believe that the Prime Minister should set up, and herself chair, a high-level National Science Council, to determine our overall science policy.

It almost beggars belief to witness senior professors actually asking for their academic freedom to be removed.

When the government's University Grants Committee was established in 1919, it was established under the Haldane Principle, named for the Liberal statesman, to protect the universities' freedom, so the UGC was initially staffed mostly by academics, whose decisions were largely autonomous. But a government rarely surrenders its control over money. It will often allow its dependent institutions some initial autonomy, but once those institutions have displaced the institutions of the free market, a government will generally exploit its monopoly powers to rule. Thus the UGC has now been supplanted by the Higher Education Funding Councils (HEFCs), and government ministers now direct the policies of the universities. In particular, they have removed the universities' economic freedoms, and the universities' fees (which are low at £3,000 a year for undergraduates) are set by the politicians. Do you suppose Harvard would be good today if its fees were capped at $5,000 a year? No wonder that, over the last twenty years, funding per student in Britain has halved, staff:student ratios have also halved, and academic salaries have fallen below those of skilled manual labourers.

ANOTHER OTHER COUNTRY

The vast post-Second World War university expansion, funded by government, was fuelled by another other country, no longer Germany but the USSR, whose launch of a sputnik in 1957 provoked a similar crisis in the UK as in the US. It seems laughable now, but between 1950 and 1970 people really feared Russian economic dominance. The Russians were Baconians and they were spending 3.7 per cent of GDP on R&D. Although much of that was for defence, it still dwarfed West Germany's 2.8 per cent, Japan's 2.8 per cent, the USA's 2.7 per cent, the UK's 2.3 per cent and France's 2.1 per cent.[9] In his 1959 essay *The Two Cultures and the Scientific Revolution* C.P. Snow wrote that the Russians were overtaking us economically because they were ruled not by arts graduates but by engineers: 'The Russians have a deeper insight into the scientific revolution than anyone else.'

So the government copied Soviet policy in science and higher education. The first sputnik was launched in 1957, and in 1958 a Treasury minute authorized the creation of eight new 'plate glass' universities with Shakespearian names, including Essex, Kent, Lancaster, Sussex, Warwick and York. The publication in 1963 of the Robbins Report on Higher Education (which was obsessed by Russia's economic, technical and social 'successes') spawned a further fourteen universities. Everyone believed in the USSR: Nye Bevan, the left's darling, predicted to the Labour Party Conference of 1959 that 'the challenge is going to come from Russia. The challenge is not going to come from the United States, West Germany or France,' and it was Harold Wilson, soon to be prime minister, who delivered his 'White Heat of the Technological Revolution' speech to the Party Conference at Scarborough in 1963:

> Those of us who have studied the formidable Soviet challenge in the education of scientists and technologists, and above all, in the ruthless application of scientific techniques in Soviet industry, know where our future lies.

The collective creed was expressed by Bill Maitland, the hero of John Osborne's *Inadmissible Evidence* (1964), whose only faith was 'in the technological revolution, the pressing, growing pressing, urgent need for more and more scientists, and more scientists, for more and more schools and universities and universities and schools, the theme of change, realistic decisions based on a highly developed and professional study of society by people who really know their subject, the overdue need for us to adapt ourselves to different conditions'.

On becoming prime minister in 1964 Harold Wilson acted on his beliefs, pouring money into science and higher education. Yet what was the consequence? Was it the 4 per cent annual growth in GDP that

Wilson predicted or was it economic disaster? Yup, you guessed it. So comprehensively did the British economy collapse that in September 1976, in return for a vast loan, the Labour government handed over its direction to the International Monetary Fund. Wilson's faith in science as a certain driver of economic growth had been exposed as nonsense.

To be fair, the Baconian experiment worked scientifically. It is often forgotten that not only was the world's first commercial nuclear electricity generating station British (Calder Hall, commissioned in 1956), so was the world's first commercial computer (sold by Ferranti in 1951), the world's first commercial jet airliner (Comet, in service in 1952) and half of the world's first and only commercial supersonic airliner (Concorde). But the Baconian experiment failed commercially in the UK, as it had in the USSR, because these advances, though having received vast government-funded R&D, were all financial disasters.

Consequently, in 1971, Shirley Williams, who had been Labour's Secretary of State for Education and Science, warned that 'for the scientists the party is over', and for the next quarter of a century successive British governments – to the rage of scientists who had grown accustomed to government science budgets growing much faster than national income – did not increase science funding much above inflation. Only after 1997, on Labour's re-accession to power, did government science budgets lurch up again, from £1.3 billion to £3 billion ten years later, under the influence of endogenous growth theory.

SCIENCE POLICIES ELSEWHERE

The history of government science policies worldwide can be easily summarized. First Britain and then the USA grew rich through laissez-faire. Many governments over the rest of the world, trying to catch up, imitated the private British and American universities and research laboratories by funding their own. Yet no one has ever shown that the government-funding of science has anywhere stimulated growth in GDP per capita.

Japan is particularly interesting. As late as 1991, inspired by British and American growth under laissez-faire, its government was funding less than 20 per cent of its R&D and, remarkably, less than half of its country's academic science – an extraordinary exception to the average OECD government, which was funding around 50 per cent of its R&D and 85 per cent of its country's academic science.[10] Yet Japan's scientific laissez-faire witnessed only decade after decade of stunning economic growth.

Only after 1990 did the Japanese economy stagger, so since then, hoping to buy its way out of trouble via science, the Japanese government has invested massively in research. But that strategy has failed. Japan Key-TEC, for example, was launched in 1985 with the equivalent of US$1.1 billion to develop biotechnology and tele-communications. A decade later, with most of the projects completed, the government's Management and Coordination Agency reported that it had achieved a return of . . . 0.5 per cent![11] Japan's problems have never lain in a shortage of science or technology. Since 1980 Japan has required reform, not government money for research.

* * *

Let us therefore note that the history shows that the funding of science by government does not stimulate economic growth. It didn't in Harold Wilson's Britain, it didn't in the USSR, and it won't in Japan. Science, therefore, cannot be the public good of Bacon's ideas.

Book 3
What is Science?

14

There is no Such Thing as
Science, There are Only Scientists

'There is no such thing as society.' Margaret Thatcher, 1983
'There is really no such thing as art. There are only artists.'
Ernst Gombrich, *The Story of Art*, 1950

During the early 1980s the Japanese government initiated a major project, the fifth-generation supercomputer. This was to be a massive but user-friendly parallel computer of 1,000 processors, with software based on logic rather than on conventional structured programming. Oh, the panic! The Japanese were going to take over the world's electronics industry!

The Japanese government's real initiative, of course, was to spend taxpayers' money on research into technology – $400 million, at least. Japan's previous technological successes had been privately funded, and its failures had been funded by the government, but the early 1980s witnessed the apotheosis of the MITI myth, and European and American governments scrambled to compete. The British government launched the Alvey programme to subsidize manufacture and research in electronics, while the European Union launched similar programmes, including ESPRIT, BRITE/EURAM, COMETT, COST, EUREKA, MONITOR, RACE, SPRINT, Telematics and VALVE.

Meanwhile, the US government, when it discovered that half of the computer chips in the F-16 fighter's fire control radar came from Japan,

SEX, SCIENCE AND PROFITS

launched its own $100 million programme, Sematech, to subsidize electronics research. But Sematech was more than a programme of subsidy. Supposing that the US electronics market had 'failed' because too many companies were – horrors – competing, Sematech was designed to integrate the different companies into a research consortium on the German nineteenth-century cartel model. That cartelization even required the 1890 Sherman Anti-Trust Act to be redrafted as the National Cooperative Research Act, thus showing how, even in the US, faith in the market will evaporate in the face of 'the other country', fears of relative economic decline and worries over defence.

And what were the fruits of these different programmes of government subsidy? Er, not many. In Britain and Europe electronics manufacturers still struggle, whereas in America the companies that joined Sematech have done no better than those that kept out of it. In the words of T.J. Rodgers, the president of Silicon Valley's Cypress Semiconductor, one of the companies that kept out of Sematech, 'consortia are formed by those who have lost'.

The different government programmes on both sides of the Atlantic yielded little because they were subsidizing yesterday's science. The governments were subsidizing research into hardware at the moment that Bill Gates *et alia* were making their advances in software. Indeed, not only did Gates *et alia* succeed without government support, but Sematech's cheerleaders in the federal government (men like Al Gore, who wanted to remould all of America's research as a government-led cartel in Sematech's image) then dedicated much of the late 1990s to restraining Microsoft, which, to their dismay, thrived in the absence of government patronage.

Governments are, of course, dreadful judges of commercial opportunities. In Japan MITI had been forced to fund the fifth-generation supercomputer itself because it could not find anyone in the commercial sector stupid enough to do so, and, in a chain reaction of hubris, the European and American governments followed suit. Yet governments continue to spend money on science and technology because they, like nearly everybody else on the planet, subscribe to Bacon's two false models, the linear and public good models. The linear model is:

Government money → academic science → technology → wealth

The public good model supposes that only governments will fund research, whether academic or technological. Let us review these two ideas.

WHENCE DOES TECHNOLOGY COME?

Everyone agrees that technology creates wealth, but the question is: whence does technology come? The first systematic attempt in recent times to answer this question came from the UK, in a book published in 1972, *Wealth from Knowledge: A Study of Innovation in Industry*. The study was led by F.R. Jevons of the University of Manchester, who, like everyone else, expected to find that advances in technology emerged from advances in basic science. In Britain, then, useful technical innovations were recognized officially with Queen's Awards for Industry, but when Jevons looked at the origins of the eighty-four innovations that had won the Award in 1966 and 1967, he found to his surprise that they were not based on science, they were based on pre-existing technology. In his own words: 'Although scientific discoveries occasionally lead to new technology, this is rare.' More generally, 'technology builds on technology'. And Jevons found that the building of technology occurs not in university laboratories but within the R&D departments of industry.

Building on Jevons's study, the economist Edwin Mansfield of the University of Pennsylvania went on to provide the most comprehensive analysis of the sources of new technology. For the years 1975 to 1985 Mansfield surveyed, seventy-six major US firms that, collectively, accounted for one-third of all sales in seven key manufacturing industries: information processing, electrical equipment, chemicals, instruments, drugs, metals and oil. And he discovered that for those seventy-six firms, 'about 11 per cent of new products and about 9 per cent of new processes could not have been developed, without substantial delay, in the absence of recent academic research'.[1]

That was a devastating study because it showed that some 90 per cent of industrial innovation arose from the in-house industrial development of pre-existing technology, not from recent academic science. Moreover, an earlier study, commissioned by the National Science Foundation, found (to the NSF's horror) exactly the same, namely that 90 per cent of industrial innovation arose in-house.[2] And the implications for the linear model are grimmer yet: though Mansfield found that around 10 per cent of industrial innovations emerged out of academic research, he also found that those innovations tended to be economically marginal, accounting for only 3 per cent of sales and 1 per cent of the savings or profits that industry made through innovation. The linear model therefore is not linear at all, in reality it needs a separate origin, a fork and lots of arrows:

academic science ↔ new technology → economic growth
↑↑↑↑↑↑↑↑↑
old technology

And the so-called linear model also requires a reverse arrow, because technology can lead as vigorously to advances in basic science as vice versa. Indeed, science has owed more over the centuries to technology than has technology to science: witness Sadi Carnot's *Reflexions sur la puissance du feu*, which transformed thermodynamics after Carnot realized that Newcomen's steam engine could not be explained by contemporary physics.

This pattern of technology feeding science continues to be repeated. Consider radioastronomy. As we know, radioastronomy was born in industry when Karl Jansky, an engineer, was tasked by Bell Labs to discover the source of the noise that was disrupting its embryonic overseas radiotelephone service. Less well known is that the young science was brought to maturity, also in the private sector, by a self-funded amateur, Grote Reber (1911–2002). Reber, a radio engineer, built the world's first radiotelescope as a parabolic dish reflector in his backyard in Chicago in 1937. It was Reber who first charted the radio sky, and until his death in 2002 at the age of ninety, Reber continued to study the sky, having moved to Tasmania because only there can very low frequency radio waves penetrate the ionosphere. And it was again in industry where radioastronomy made perhaps its greatest discovery when Arno Penzias (b. 1933) and Robert Wilson (b. 1936) of Bell Labs discovered the cosmic microwave background radiation, thus making the observations that underpin the Big Bang. They won Nobel prizes in 1978.

That same pattern, of industry and the private sector conceiving and fostering a new science, can be seen in biology. Even mighty molecular biology arose from privately funded applied science. How do we know that DNA, a sugar, is the molecule of inheritance? Because Oswald Avery (1877–1955), a doctor, was studying pneumonia at the privately funded Rockefeller Institute. During the 1940s pneumonia was still a major killer, and Avery was studying *pneumococcus*, one of the bacteria that causes it. *Pneumococcus* comes in two forms, encapsulated and non-encapsulated, and Avery discovered that if he injected DNA from the encapsulated into the non-capsulated he transformed it permanently into the encapsulated. Thus, out of a piece of privately funded applied research, did Avery discover that the sugar DNA (not proteins as was then thought) regulated inheritance. Once Avery had revealed the importance of DNA, the discovery of its structure became inevitable and thus merely a race (between Linus Pauling, James Watson and Francis Crick, and Rosalind Franklin and Maurice Wilkins). Molecular biology has been a race ever since, as teams chase each other towards predictable goals.

The history of science and technology continues to show that they fertilize each other, which is why so many Nobel prizes continue to be won by industrialists. The Nobel prize for the discovery of the transistor, for example, went to W.B. Shockley and his colleagues from Bell Telephone Labs (again!) while the prize for the discovery of the polymerase chain reaction (of *Jurassic Park* fame) went to Kary Mullis of the Cetus Corporation. Such examples can be repeated ad nauseam.

At a practical level, science cannot progress without advances in technology. But for the development during the nineteenth century of really tight vacuum tubes, for example, there would have been few of the experimental advances that yielded radioactivity, atomic structures or quantum physics. Thus we see that, at the practical as well as at the conceptual level, science has long owed more to technology than vice versa.

IS SCIENCE A PUBLIC GOOD?

OK, so we've largely disproved the linear model, but basic science surely plays *some* role, no matter how small, in economic growth: we can all think of technological developments that have flowed out of basic science. So let us ask if the market would fund enough of it. The economists say no.

Economists argue that science is a public good. Anyone can freely read the scientific journals, so why should a private individual or company invest in research when they will preferentially benefit others, including competitors, enemies and the unborn? An influential study that apparently confirmed the folly of investing in science was published by two Japanese economists, Hiroyuki Odagiri and Naoki Murakimi of the University of Kushiro. Odagiri and Murakimi surveyed the ten major Japanese pharmaceutical companies that, in 1981, together enjoyed $13 billion sales. Those companies performed R&D, and Odagiri and Murakimi found that each individual company could expect, on average, an annual return on its investment in R&D of 19 per cent. Not bad. But each company's competitors benefited more. By exploiting the advances made by any one company (published as patents and papers or sold as products) the other nine companies could subsequently create new drugs that, together, provided the equivalent of a 33 per cent annual return on the first company's research: other people's prizes were worth nearly twice as much as the originators'. As Odagiri and Murakimi concluded:

> Thus the private returns are argued to be smaller than the social returns, and government support is called for to make up for the reduced private incentive to research.
>
> H. Odagiri and N. Murakimi, 'Private and Quasi-social Rates of
> Return on Pharmaceutical R&D in Japan', in *Research Policy*,
> 1992, 21:335–45

So, the economists say, private companies will not fund science because others (the 'social' return) will benefit preferentially. Therefore governments must fund science – especially the pure or 'basic' science that the economists still suppose to be the basis of applied science. The classic expression of this belief was made by the celebrated US economist Paul Romer:

> Research has positive external effects. It raises the productivity of all future individuals who do research, but because this benefit is non-excludable [ie, enjoyed predominantly by others], it is not reflected at all in the market price [ie, which is too low to incentivize research].
>
> P. Romer, 'Endogenous Technical Change' in *Journal of Political Economy*, 1990, 98:S71–S102

As another prominent US economist, William Baumol of the universities of Princeton and New York, writes:

> . . . government's financing of basic research, whose spillovers are particularly large [ie, whose benefits are enjoyed predominantly by others], is quite appropriate.
>
> W. Baumol, *The Free-Market Innovation Machine*, 2002, p.141

Yet companies do, in practice, fund science, including pure or basic science, extensively. Repeated surveys, even those of the Science Policy Research Unit at the University of Sussex (a leading UK lobbyist for government money for science), show that some 7 per cent of all industrial R&D worldwide is spent on pure science.[3] A recent study of British industrial R&D (produced again by SPRU at Sussex) revealed that some UK companies, including ICI, SmithKline, Beecham, Wellcome and AEA Technology, each published more than 2,000 papers between 1981 and 1994, exceeding the output of medium-sized universities.[4]

Indeed, the quality of science performed in industry is startling. The *Times Higher Education Supplement*'s 2005 research survey showed that Harvard University is the institution whose science papers are the most cited globally (20.6 citations per paper on average), but coming in second was IBM (18.9), outranking all other universities and research bodies. A survey by the Institute of Scientific Information revealed that, of the seven producers of the highest cited papers in biology, two were private profit-making companies, Genentech and Chiron, whereas another was a wholly private charity, the Howard Hughes Foundation, and three others were private charities, the Salk Institute, the Cold Spring Harbor Laboratory and the Whitehead Institute.[5] Though those three do now receive government research grants, only one of the seven institutions, the Institute de Chimie Biologique in Strasburg, France, is wholly government-funded – so much for the irreplaceability of state funding for high-quality pure science.

Companies fund pure science because, contrary to economic theory, they find it highly profitable. Two major US surveys, one by Edwin Mansfield and the other by Zvi Griliches of Harvard, have shown that the more basic science a company performs, the more likely it is to grow and to outperform its competitors. Mansfield studied sixteen major American oil and chemical companies for the years 1960–76 and showed that all those firms invested in pure science: crucially, the more a firm invested in basic science, the greater its productivity grew.[6] Griliches, in a comprehensive study of 911 large American companies for the years 1966–77, showed that the companies that engaged in basic research consistently outperformed those that neglected it: the more basic research a company performed, the greater its profits.[7]

Basic research, therefore, turns out to be profitable, and the companies that fund it grow to command the market. Why? Because the economic theory is wrong.

SOME REAL ECONOMICS OF SCIENCE

Let us consider a company that does basic science in the hope of producing a new product. The company could employ a team of scientists, lock them away in a laboratory, and tell them not to emerge until they have made some discoveries. Since scientists love doing unfettered science, they will love their new task. They will seize on interesting problems, make discoveries (yummie, scientists *love* making discoveries), write papers, present their results at international meetings and generally have fun. This is not fiction: one company that, during the 1960s, based an R&D strategy on such an investment in basic science was SmithKline.

SmithKline knew that stomach ulcers are caused by acid. And the company also knew that the secretion of stomach acid is regulated by histamine, as are the symptoms of hay fever. But whereas the anti-histamines then available could treat hay fever, they could not block the production of stomach acidity. So SmithKline charged a team of scientists, led by James Black, to explore the science of histamine and stomach acidity. Black and his team had a ball. They guessed that in the stomach there must be a second class of histamine receptor, histamine 2, which, unlike the hay fever receptor, would be blocked only by novel antihistamines. After the expenditure of much SmithKline money they found both their receptor and their specific histamine 2 blocker, Tagamet. That discovery helped propel Black to his Nobel prize in 1988, but did it make SmithKline much profit? Er, no. The company that did make the profits, with Zantac, was Glaxo. Zantac is the drug of choice for ulcers, and it has made Glaxo well over $10 billion in profit.

The head of Glaxo's research team was Dr David Jack. Jack was a copy-cat. One night he attended a lecture of Black's, and he returned to Glaxo the next day determined to produce a more potent derivative.

Copying is easier than original research, and Jack soon came up with Zantac. In Jack's own words: 'It was a straight piece of medicinal chemistry because the original thinking had been done by Jim Black. It does however show you something very important. The second prize in this business can be bigger than the first.'[8]

So, are the economists not correct? Is science not a public good, most of whose benefits go not to those who make the discoveries but to those who only copy them? Black at SmithKline might have done the original research, but were not the ideas – and the bulk of the profits – snapped up by Jack for Glaxo?

Ah, but could *you* have done what Jack did? David Jack may have accessed Black's work at an evening lecture, but could you have? If you had swept off the streets into Black's lecture hall, would you have rushed off the next day to copy Tagamet? Or would you have gone home to a reality TV show and wondered what all that incomprehensible rubbish of Professor Black's was about?

It is, quite simply, a myth that published science is freely available to the public. The number of people who, on hearing Black speak, could have gone off and copied him was tiny. Indeed, the number of people who would have even been invited to listen to him was tiny. Research is published in specialist journals or delivered at specialized lectures to which only specialists have access. There are no public or social returns to science, only specialist or (to be more accurate) collegiate ones.

COPYING IS GOOD BECAUSE IT IS THE BASIS OF SCIENCE

We can now begin to understand why companies invest in pure science. As the economists Wesley Cohen and Daniel Leventhal noted in their 1989 paper 'Innovation and Learning: The Two Faces of R&D',[9] companies do research as much to learn from others as to make their own discoveries. Science is a vast enterprise, and no single company can begin even to anticipate all the discoveries being made daily all across the world. No matter how big a company, most of the discoveries that matter to it will be made elsewhere. Consequently, so-called 'second-mover' advantages often outweigh 'first-mover' advantages, and it is often better to be an opportunist than a pioneer: it is generally better to keep up with all the world's research – to develop opportunities as they arise in others' labs – than to try continuously to invent the wheel oneself.

Some companies even boast of their copy-cat priorities. The research prospectus of Mitsui Pharmaceuticals of Japan states as its first priority: 'The exhaustive survey of the scientific literature, collection and analysis of various information, establishment of R&D targets, and coordination between various research groups.' Only secondly does Mitsui prioritize: 'The chemical synthesis of new compounds . . . and the exploration of new biologically active substances from natural sources.'

But, and this is the big but, who is going to do the exhaustive surveying of the literature and the collecting of information? The only people so qualified are competent researchers, because only they can actually read and understand the papers. It is very hard (because so much science is tacit, see later) to understand new science, and the only people so capable are active researchers in the field. So if a company is to access other companies' work, it will succeed only if it employs good scientists, because the better the scientists, the better they will be at importing knowledge. And the best scientists will demand the freedom to follow their own first-mover research, because the best scientists (like the best artists) know that only they – not the suits on the board – know how best to explore the limits of knowledge.

Thus did David Jack of Glaxo negotiate 'final control of everything the research department did'.

No such negotiations were necessary at Genentech, where the legendary Robert Swanson had already understood that he would attract the best scientists only if he allowed them to research freely and to publish as freely. But such negotiations were necessary at Du Pont. Du Pont has been doing R&D for over a century, and over that time it has experimented with different ways of doing R&D, sometimes providing its scientists with the freedom to research and publish at will, sometimes binding scientists to corporate goals and to secrecy. And it found that when scientists were discouraged from publishing or from researching freely, their productivity fell, crippled by low morale and resignations. Du Pont has learned to provide its scientists with freedom.[10]

Which is why companies spend some 7 per cent of their R&D budgets on pure science, because without funding pure science, companies would not retain the scientists to scour the journals, trawl the meetings and talk the talks that yield others' first-mover nuggets that spawn the in-house second-mover profits. But to retain good scientists, a company needs to provide them with laboratories, money and the freedom to publish, much as a company pays consultants' fees: in effect, scientists are in-house consultants. Their real value to the company arises not so much from the research they do in-house (nice though that may be) as from the information they import.

Consequently, the copying of others' research is vastly expensive. It is simply a myth that science is freely available to be copied for nothing. When Jack and his research colleagues at Glaxo were copying Tagamet to produce Zantac, Glaxo's profits fell from £87 million (1977) to £60 million (1980) as research costs rose from £17 million (1976) to £40 million (1981).[11]

A number of systematic surveys have confirmed the high costs of copying. When Edwin Mansfield examined forty-eight products that, during the 1970s, had been copied by companies in the chemicals, drugs, electronics and machinery industries of New England, he found that the costs of copying were, on average, 65 per cent of the costs of original invention.[12] Copying, moreover, is time-consuming, taking on average 70 per cent of the time of original invention. Copying is expensive and time-consuming because each copier has to rediscover for themselves the tacit information embedded within the innovation. Mansfield found that some inventions actually cost more to copy than they had to invent because the copiers had had to acquire so much tacit knowledge. What is this tacit knowledge?

TACIT KNOWLEDGE

The centrality of tacit knowledge was discovered by Robert Boyle (1627–91) at the very beginning of the enterprise we call science. Boyle was a great researcher, and like many great researchers of his day he was a rich man, which is why he was known as the 'Father of Chemistry and the son of the Earl of Cork'. Today Boyle is still remembered for Boyle's Law (which states that, on being squeezed, a gas contracts predictably), yet his immeasurable contribution to science was not the research he did but the rules he created. To a considerable degree, it was Boyle who helped institutionalize the way we all now do science, and in his books, including *The Sceptical Chymist* (1661) and *A Free Enquiry into the Vulgarly Received Notion of Nature* (1686) he described how scientists should behave.

One of Boyle's rules was that a research paper should provide sufficient details for any other researcher to reproduce the experiments – that is, a paper had to have a 'methods' section as well as a 'results' section. In demanding such transparent publication, Boyle was reforming an unhelpful aspect of the science of his day, namely that of discreet publication. To establish priority, the scientists of his era would publish, but often only secretly. They did not want others to benefit. So some scientists, having dated the report of a discovery, would seal and deposit it with a college or lawyer, to open it only to dispute priority with a later competitive publication. Others would publish in code or in anagrams: Robert Hooke (1635–1703) published his law of elasticity as *ceiiinossttuu*, which transcribed into *ut tensio sic vis* (stress is proportional to strain; anyone who thinks that Newcomen developed his steam engine because he could break Hooke's Latin code has a lot of faith in the cryptic classical language skills of a near illiterate Devon yokel).

Yet, ironically, nobody could reproduce Robert Boyle's own experiments, no matter how carefully he described them. All over seventeenth-century Europe, scientists tried and failed to repeat Boyle's work. Eventually people realized that, to reproduce his experiments, they had to visit Boyle in person and handle and dismantle and operate his air pumps themselves. Such visits to (or training in) the labs of other scientists remain essential nowadays, and today we appreciate that much scientific knowledge is 'tacit' (from *tacitus*, Latin for silent) and that published work, or work described orally at conferences, can guide only those people who already possess the relevant tacit knowledge in the field.

Consequently, the copying of other people's technology is not cheap but, rather, expensive, because the most difficult part of the copying is the laborious repetition of the inventors' acts by which to acquire their tacit knowledge. And not even specialists can easily acquire the tacit knowledge behind the latest discovery.

Edwin Mansfield may have reported that copying costs 65 per cent of the original discovery, but even that figure of 65 per cent does not cover the full costs of copying; it covers only the marginal or tacit ones of the actual copying. That is because copiers first have to discover, by systematically reading the research papers and patents, attending the meetings and sustaining the collaborations, that there is a new discovery or patent or process or product to copy and, as the Nobel economics laureate George Stigler showed in his 1961 paper 'The Economics of Information', acquiring that knowledge is horribly expensive.[13]

And who is going to acquire the information? Why, only competent scientists, who can maintain their competence only if their own first-mover experiments are properly funded, because only if their own competence is maintained will they retain the capacity to judge the opportunities and value of others' innovations. Copying someone else's discovery, therefore, incurs three separate costs: the actual or 'marginal' costs of copying of it; paying the scientists to discover its existence; and funding the company scientists' research to retain their tacit research competence. When those costs are summed they add on average to the costs of the original invention. Thus we see that it is not cheaper to copy than to invent, the two costs balance.

Isn't that extraordinary! Isn't it amazing that, out of all the possibilities, the evidence shows that, in the real world, the costs of innovation and of copying should balance so exactly! Er, no, actually it is inevitable. Of course, an economy is going to develop such that the costs of innovation and of copying balance, because entrepreneurs will titrate their investment between the two activities appropriately. The vast funds spent by private industry on civil R&D ($181 billion in the US alone in 2000, which dwarfs the government's $72 billion and the foundations' $12 billion[14]) should have signalled to economists that science is far from being a public good.

WHY DO SCIENTISTS PUBLISH?

Around 1600 a young London-based obstetrician called Peter Chamberlen invented the obstetric forceps, and for over a century he, his younger brother, his younger brother's son and that son's son (all obstetricians) kept the invention a secret. Rich women, knowing that the Chamberlens were the best obstetricians in Europe, engaged them to deliver their babies, but the price those women paid (apart from handsome fees) was to be blindfolded and trapped alone with the Chamberlens in a locked room during labour so that no one could discover the secret of the forceps. That emerged only during the 1720s when the last Chamberlen, having retired rich but childless, finally divulged it. Which raises a key question: why should anyone publish discoveries when that publication will preferentially benefit other people?

Sometimes the answer is relatively trivial. Companies, for example, publish to raise their credibility with regulators, patent agents and other stakeholders, and to benchmark the quality of their science (such benchmarking not only aids the recruitment of new scientists but also reassures the firms' management that their scientists' R&D is competitive). Yet that sort of utilitarian publication is not central to science; what is central is that all scientists, wherever they work, are desperate to publish. Why? This is so important – it goes to the heart of science – that I need to make a detour into evolution by sexual selection.

Evolution by sexual selection

Charles Darwin is remembered for describing evolution by natural selection, but Darwin also showed that sexual selection is as important to evolution. Darwin observed the peacock's tail, and he noted that such a tail could only hinder the physical survival of the bird – it would snag on branches and provide predators with a clawhold. Yet Darwin understood that for the peacock, as for all males, physical survival is not enough, he has to find a female with whom to breed. Since peahens like good tail, and since peahens can choose between males, the sexually successful male grows good tail. He advertises his fitness.

Humans, too, advertise their sexual fitness. They do it by competing for esteem. Businesspeople, for example, advertise their fitness by making money, not for its own sake but as a route to esteem. Aristotle Onasis said, 'if women didn't exist, all the money in the world would have no meaning'. The evidence that money is actually only a route to esteem has been shown by testing people in psychology laboratories. The Harvard scientist Terence Burnham reported that when subjects are offered small sums of money, they accept them – unless they see others being offered more. Then they reject the original small sums.[15] People are not formally rational *Homo economocii*, they are not interested in absolute wealth, they are interested only in relative wealth – esteem.

Which is why some people, including artists, politicians, scientists or fashion models, bypass money altogether and advertise their fitness directly by their eminence in the arts, politics and the sciences or on the catwalk. Scientists, therefore, are amongst the people who seek esteem not by money but by creative distinction. This is what Geoffrey Miller, the psychologist who studies sexual selection, wrote about science in his 2000 book *The Mating Mind*:

> Science is a set of social institutions for channelling our sexually selected instincts for ideological display . . . The rules award social status for proposing good theories and gathering good data.

This explains, of course, why scientists are often poorly paid – they're not in it for the money. Indeed, they pay to succeed – so desperate are scientists to publish, they will spend some of their grant money for the privilege. Many of the top academic journals charge their authors in excess of $1,000 for each paper they publish, which even requires those journals to formally label each paper an advertisement. And we are moving to a new world of electronic publication, where papers will be freely available on the web. In that new world readers will no longer pay to read journals: instead, authors will pay to publish. Indeed that world has already arrived in the form of electronic open-access journals such as *Public Library of Science Biology*.

That new world is being promoted by governments, industrialists and learned societies to reduce the costs of information, but the fundamental drive is coming from the scientists themselves. Since science is a community, the writers and readers of journals are the same people, so the question of 'free' journals can be seen as simply logistic; is it more efficient to shift payment from readers to authors? It is only snobbery that ensures that high-status publications from high-status labs are still reserved for the paper-published journals.

The economics of science publishing, in short, are those of vanity publishing, so if the discipline we call economics is the study of incentives, then scientific publication cannot be the problem the economists fear, because it is the incentive to do science. Conventional economic thinking is wrong.

Sex and scientific creativity

Science is a creative activity and sexual selection also explains that. Consider the humpback whale. Both the Indian and Pacific Oceans support humpback whales, but the two populations are normally kept apart by the bulk of Australia. In 1996, however, Michael Noad and his colleagues from the University of Sydney noted that two males from the Indian Ocean had, during the mating season, joined a group of eighty-two humpbacks that were migrating along the Pacific side of Australia.

During the mating season male humpbacks sing to attract mates. All the males in a population sing the same song, but each population's song changes gradually over time. Different populations of humpbacks therefore sing different songs, so the two Indian Ocean males were singing a different song from their new Pacific friends when they joined them in 1996. Yet by 1998 all the Pacific whales were singing the newcomers' Indian song.

Female humpbacks, it transpires, are attracted to novelty, and the Pacific males switched to the Indian tune when they found their womenfolk being unduly attracted to it.[16] Equally, humans are attracted to novelty. Our head hair, for example, grows continuously to create novelty by allowing us to cut and shape it into ever-changing fashions. Similarly, we humans lost our body hair about 70,000 years ago to provide a canvass to decorate with clothes, body painting, tattoos, jewellery, and make-up. The canvas that is the naked body encourages novelty in clothes and decoration, which is why the young are forever adorning themselves in novel fashions, the better to attract mates. Human cleverness and creativity thus appear to have evolved because those who were clever and creative enough to joke, paint, sing and groove to the beat on the dance floor got more sex than the bores, just as the whale brain apparently evolved its size in part because only the most original singers got the girls.

Leaders, too, get more sex than do the peons, and no one observing the endlessly shifting alliances and stratagems of our fellow apes in their ceaseless search for sexual monopolies and power need doubt where much of the origins of primate originality lie. In more formal language, the human brain evolved to exploit the seductive, competitive and reciprocal rewards of communal living. And in a process of 'runaway co-evolution' we humans, both male and female, have invested ever more in creativity. Thus did sexual selection give us cleverness and creativity – and it thus did it give us science, that cleverest and most creative of activities.

WHY DO SCIENTISTS CLUSTER?

OK, so we've outlined the evolutionary roots of scientific creativity. And we've outlined the evolutionary roots of scientists' desire to publish. But we have not explained why scientists publish material that benefits their rivals. Surely the scientists of Robert Boyle's day were more rational than today's when they published only the results of their findings (and kept their methods secret) or when they did not publish at all but merely lodged their findings (to release them only to establish priority if the face of a later competitive publication)? To address this question, let me note that scientists not only publish papers but they also cluster in conferences, literally to broadcast their discoveries to their rivals. So, if we can understand why scientists cluster, we might also understand why they publish generally. One clue is provided by the related phenomenon of industrial clusters.

Since 1990, when Michael Porter published his *Competitive Advantage of Nations*, it has been fashionable to talk about clusters, and people have looked to Silicon Valley, Route 128 or to their local science park for miracles of technology. These clusters contain companies, often direct competitors, congregated together. But why should competitor firms cluster?

Some clusters have a long history. They are called cities. Cities are the centres of technological growth because, in their large size and density of population, cities optimize competition, demand and communication. And cities can promote cross-fertilisation between different industries. The classic example was the printing press, which emerged in Mainz when Gutenberg integrated four distinct local technologies: the printing of large sheets of cloth by wooden blocks; the printing of playing cards by movable wooden type faces; metal casting (for moulding resilient type faces); and wine presses. Modern industry throws up myriad examples of cross-industrial fertilization: carbon composite material, for example, is lighter than aluminium, yet as strong, so it provides excellent fuselages, wings and other parts of aeroplanes. But carbon composite needs to be kept cool during manufacture, and the aerospace manufacturer Northrop found it could replace aluminium with carbon composite only after some friendly refrigeration experts from Sara Lee, the frozen food people, had explained how to refrigerate an entire factory. Such inter-industrial links are fostered by proximity.

Industrial clusters, however, are different from these city-based inter-industrial partnerships: clusters congregate competitors, cheek by jowl. A person might suppose that competitors would want to distance themselves from each other, but trades have often clustered. In London, jewellers have long clustered on Hatton Gardens, and other trades have clustered around other streets. Trades have sometimes clustered on whole cities. Sheffield in England, like Solingen in Germany, has traditionally specialized in cutlery, and there many other examples, including Detroit, motown.

The nineteenth-century economist Alfred Marshall described the clustering of industrial competitors as providing 'external economies of scale'. Those external economies might include the sharing of specialist suppliers as well as access to specialist labour, specialist consultants and other industry-particular specialists that no single company could support. But there must be more to clustering then those externals because, long after industries have outgrown their original geography, they can remain clustered. New York, for example, developed as a financial centre to serve the needs of a port with good access to the

Midwest yet, though modern electronics have rendered the geography redundant, New York remains a financial centre.

Shared knowledge emerges as the key. Surprisingly perhaps, competitor companies share information. In a survey of eleven American steel companies, Eric von Hippel of MIT's Sloan School of Management found that ten of them regularly swapped proprietary information with their rivals.[17] In an international survey of 102 firms Thomas Allen, also of MIT's Sloan, found that no fewer than 23 per cent of their important innovations came from swapping information with rivals: 'Managers approached apparently competing firms in other countries directly and were provided with surprisingly free access to their technology.'[18] In a cross-disciplinary study, Louis Galambos, Professor of History at Johns Hopkins University, and Jeffrey Sturchio, vice president of external affairs, Merck Sharp & Dohme, have shown how pharmaceutical companies – though intense competitors – will also share knowledge.[19]

Practical businesspeople have long known that rivals share information. It was the president of the Western Electronics Manufacturers' Association in the US who reported that competitors 'share the problems and experiences they have had.'[20] Some companies even boast of being good sharers:

> Conventional business wisdom says: Never let the competition know what you're doing. But at Novell, we believe the secret of success is to share your secrets. So we established the Novell Labs Programme to openly share our networking software technology with other companies.
>
> Advertisement in *The Economist*, 21 September 1991

We have here an important insight into why governments need not fund science. One standard economists' argument is that the research of private industry is inefficient because companies will duplicate their research and not share advances; but in practice company scientists do share their results and avoid duplication. Government money is not necessary for knowledge to be shared or unduplicated.

Companies share knowledge for a number of reasons, but the most important is that those companies that share knowledge will outperform those companies that do not, because such sharing widens a company's knowledge base and thus its opportunities. And the firms with which to share are those with the same specialist knowledge – rivals. But sharing information with rivals requires opportunity and, even more importantly, it depends on trust. Opportunity and trust are facilitated by proximity, so companies cluster to get to know each other. Nothing propinqs like propinquity.

And we have here, therefore, one explanation for why academic scientists, too, congregate – not to *give* information away but to *trade* it. Though researchers compete, they also benefit from trading knowledge with their competitors, because non-sharers get left behind. So the scientific conference appears to be a forum for donating information – for giving it away publicly – but actually it is a trading floor. Ignore the formal plenary lectures: they are obsolete, because they recount merely what has already been published. Ignore the unsolicited material proffered by scientists as oral or poster communications: it is dubious, because it has not been peer reviewed. In fact, almost all the formal aspects of conferences are irrelevant.

What *really* happens at conferences is that people talk to each other, over coffee, lunch and dinner. Conferences are treasured for the private discussions that researchers hold with each other, where they share unpublished and unpublishable (tacit) material with their peers. That is why scientists meet in the flesh. Which they must do. In his 2007 autobiography *How to Avoid Boring People*, James Watson, of great DNA fame, advised scientists to 'stay in close contact with your intellectual competitors'. Otherwise they will be bypassed like Rosalind Franklin (of modest DNA fame) or the steel maker Robert Mushet (see above).

THE DANGERS OF TRADE

If conferences are actually trading floors, where researchers trade tacit and discreet information, then conferences should adopt similar institutional mechanisms to market trading floors, because all traders – whether of knowledge or goods – face the same problem, namely trust or *caveat emptor*. This is especially important in science because knowledge, being largely tacit, is so hard to verify. If you discuss your experiments with someone, trusting that, by sharing some of your findings with them, they'll reciprocate, you rely on them to tell you honestly what they've found. Otherwise you'll been taken for a ride.

Consider commerce. In commerce there are 'spot' trades – those that take place on the spot. In a shop I engage in spot trading. I exchange money for something I like, on the spot. A spot trade is therefore low-risk because a shopkeeper is hardly likely to take my money and abscond.

But most trades in an advanced economy are 'risk' trades because an advanced economy is based on long-term investments. An advanced economy requires investment in people, relationships, companies, R&D and other forms of capital, and such investment takes time to mature. Investment is therefore a long-term, not a spot trade, and it is risky because it relies on people keeping their promises. An advanced economy, in short, is based upon millions of people – millions of strangers – trusting each other to keep promises. Indeed, the very word 'credit' comes from the Latin *credere*, meaning to believe, and it shows how trade depends on trust.

But rational people shouldn't keep their promises. Adam Smith noted that in a rational world no company would survive, because no rational employee would honour their contract to work. The moment the boss was out of sight, rational employees would shirk. Equally, in a society constituted solely of rational people, everybody would cheat on their contracts and no risk contract would be sustainable. In the words of Professor Douglass North, a strictly rational world 'would be a jungle, and no society would be viable'.

Today's Third World can be like that. When I was a boy I was told that the Third World was poor because it lacked entrepreneurs, yet the Third World pullulates with entrepreneurs. Consider the airports. A traveller at a Third World airport is besieged by entrepreneurs offering a taxi ride into town or a hotel room. But neither the taxi driver nor the hotel room will inspire confidence. The Third World is poor not because it lacks entrepreneurs but because it is not 'trustworthy', so businesspeople avoid it.

Thomas Hobbes thought that everyone was like today's Third Worlders. Because he believed that humans were rational, not trustable, he believed in tyranny. Life for Stone Age people had been, Hobbes said, 'solitary, poor, nasty, brutish and short' because people would not honour their contracts: 'Covenants without the sword are but words.' The Stone Ages, Hobbes said, were therefore 'a war of all against all', which was arrested only by the rule of those nice Bronze Age tyrants who forced people to honour their contracts.

And when we look at the real world we find that some employees are indeed rational and they do indeed shirk, which is why we have evolved institutional countermeasures. Companies, for example, police their workforces. When Professor North and his colleague John Wallis examined the structure of the US economy in 1970 they found that nearly half of US GNP was spent on supervision and coordination

– only a small fraction of the US's expenditure was dedicated to the actual business of manufacture or the supply of services.[21] After they'd enumerated all the industrial supervisors, clerks, managers, executives, policemen, lawyers and accountants in the US, Wallis and North had accounted for nearly half of US GNP.

But *quis custodiet ipsos custodes*? How we can we trust the supervisors, accountants and lawyers to themselves be honest? And even if the supervisors, accountants and lawyers could be trusted, a world where everyone else was a rational cheat would still be a jungle. Market players can so easily conceal infractions from outsiders that external policing will always be feeble. Consequently, it transpires that in practice most supervisors, accountants and lawyers are largely marginal because the overwhelming proportion of trades are internally policed. Most people, in short, are trustable, not rational, and they honour their contracts. As Rousseau said, it is the law that is written in the heart that counts. Or as Henry Luce, the founder of *Time*, *Fortune* and *Life* magazines, said: 'Service is what the typical American businessman would do his best to render even if there wasn't a cop or preacher in sight.'[22]

The sociologist Stewart Macaulay, when surveying business practices in the real world, found the internal policing of contracts to be common. He described, for example, a manufacturer of packaging materials who did not create legally binding contracts for two-thirds of his orders.[23] Macaulay found mutual trust between suppliers and customers to be common, the trust being based on long-term relationships between suppliers and customers and on the desire to protect reputation. Mutual honour pays, of course, because it saves on lawyers' fees and other transaction costs. Rousseau thus confirmed that Hobbes was wrong – we can today, thanks to trust, sustain myriad risk trades within a modern economy in the absence of tyrants or Leviathans.

Equally, scientists can sustain the necessary trust by which to trade information at conferences. So, where does trust come from? Because trust lies at the heart of science, we have to take another detour by which to understand it, this time trespassing not into sexual evolution but into game theory.

GAME THEORY

In Stanley Kubrick's film *Dr Strangelove* the American president summons the Soviet ambassador to confess that, accidentally, an American bomber has been sent to nuke Russia but, hey, these things happen. Yet the ambassador retorts that the matter is truly grave because, on dropping its bomb, the American plane will activate the Russians' secret Doomsday Machine, which will destroy all life on earth. Dr Strangelove is outraged: 'The whole point of having a Doomsday Machine is lost if you keep it a secret!' Which, of course, was true, because if the US pilots had known they would destroy all life on earth, they would have returned home with their payload undropped.

Dr Strangelove was based largely on John von Neumann, the Hungarian economist who invented game theory. Game theory is now the lingua franca of the social sciences, uniting economics, biology, psychology and sociology in a common, powerful methodology: no one may call themselves educated who does not understand (non) zero sum (non) cooperative games. Yet game theory is also, at one level, trivial, having emerged by mathematicians having fun. The theory emerged by the purest of intellectual speculation, too pure even for Francis Bacon. Bacon may have believed in his linear model, but for him science was always to be pursued for power, and in his *Advancement of Learning* Bacon had maintained that:

> Knowledge may not be as a courtesan, for pleasure and vanity only, or as a bond-woman, to acquire and gain for her master's use; but as a spouse, for generation, fruit and comfort.

But game theory indeed emerged as a courtesan, for pleasure and vanity only, because mathematicians love to play games such as chess, and it was while playing such games that they began to theorize as to what a game actually was. In the *Republic* Plato had noted that even criminals had to cooperate, and he asked how 'robbers or thieves or evildoers could act if they injured each other'. The genius of game theory is that it mathematizes human relationships by quantitative analysis. Complex human relationships cannot, of course, be fully reduced to algebra, but such reductionism can be insightful. As John Williams wrote in his 1954 *Compleat Strategyst*: 'The invention of deliberately oversimplified theories is one of the major techniques of science, particularly of the "exact" sciences . . . simplified games may prove to be useful models for more complicated conflicts.'

Plato had asked how robbers could cooperate. One answer can be provided by the well-known game called the Prisoner's Dilemma. Imagine that two robbers have been caught with the proceeds of a theft. They are placed in separate cells from which they cannot communicate with each other. The police then interrogate each one, inviting them to confess to the robbery. The punishment for robbery is five years' imprisonment, but for the lesser crime of handling stolen property it is only six months. A confession (but only one that incriminates the other) is rewarded with release.

If both deny any involvement in the robbery, each will be sent down for six months for the lesser offence of handling stolen property: the police can prove that each has some of the property, but cannot prove that either stole it. If both confess, they each goes down for five years. But if one confesses and the other denies, then the confessor is rewarded for shopping his mate and he gets off scot-free. The denier goes down for five years.

The dilemma arises because it is obviously good for both of them if they both deny (six months apiece), yet it is best for each of them if he or she individually confesses (scot-free) while the other denies (five years). So both confess in the hope the other will deny, while neither denies for fear the other will confess. Both therefore go down for five years when, if they could only have trusted each other to deny, they could have got off with six months apiece.

The Prisoner's Dilemma was invented in 1945, and for many years it was believed to 'prove' that cooperation was impossible. But in 1984 the political scientist Robert Axelrod showed in his book *The Evolution of Cooperation* that the lesson of the game changes if people play it more than once. If real people under laboratory conditions pretend to be the robbers, playing the game again and again, they generally start off by confessing (the uncooperative move) but occasionally they deny (the cooperative move) to see what happens. Very quickly, the other player learns to cooperate as well, and soon both might be cooperating nearly all the time. As Plutarch said 2,000 years ago in his *Moralia*: 'Evidence of trust begets trust, and love is reciprocated by love.'

This game teaches, therefore, that if the players view it, *the game* (i.e., life) as their 'opponent', and learn to cooperate together as a team, they do best. As von Neumann said of two-person games, 'defeat is inevitable if you aim to win rather than avoid losing', so the game establishes that selfish people do well to cooperate. One good strategy is 'tit-for-tat', under which players routinely cooperate, playing a non-cooperative move only as a one-off punishment for another player's non-cooperation; having punished, players then forgive and forget.

Hobbes had feared that people, being purely selfish, would not cooperate, but game theory shows how even purely selfish people –

especially purely selfish people – will learn to act virtuously. As will their children. When successive generations have repeatedly to learn the same lesson, Nature ensures that the behaviour gets internalized as an instinct. So we humans have acquired instincts for guilt, shame, fairness, honour, generosity and the other emotions that facilitate tit-for-tat and other optimal game theory tactics.

One of the stongest game theory emotions is trust, and using a technique known as functional Magnetic Resonance Imaging (fMRI) it can even be visualized. The anthropologist James Rilling of Emory University, Georgia, showed that when people trust each other they activate particular parts of the brain. In his paper, 'A Neural Basis for Social Cooperation', Rilling showed that the brain's trust centres are located in the recently evolved parts, the parts that define us as higher primates.[24] It is trust that makes us human.

Nonetheless, we retain our selfish genes, we are still competitive and greedy and tempted to cheat, so our virtuous instincts need to be reinforced to prevent free-riding; which is why we humans, in our markets, create institutions that reinforce virtue. Those institutions have mechanisms for detecting and punishing misdeeds, and such institutions are as old as markets themselves. In his 1990 book *The Enterprise of Law: Justice without the State* the economist Bruce Benson showed, for example, how the medieval merchants of Europe spontaneously organized their own legal systems. The states of the day provided little commercial law, so the merchants created their own legal codes and courts before which misdoers were arraigned and punished. The merchants' courts, not possessing coercive powers, possessed only one minor sanction (fines) and one major sanction (ostracism), but the threat of expulsion from the club of traders provided sufficient discipline to ensure a high degree of honesty.

Equally, in 1566 Thomas Gresham (of Gresham's Law) created the Royal Exchange as London's trading floor. To make sure that his market worked, Gresham restricted admission only to people known to be trustworthy, namely to his fellow Freemasons and to his fellow members of the Mercers Livery Company. 'My word is my bond' invokes few transactions costs.

We see, therefore, that there are no free markets, not even in commerce, and that in practice markets are restricted to players whose trustworthy behaviour can be accredited. We hardly 'see' the institutions of market trust any more because we are all now institutionalized from birth, at home and at school, to be trustable traders. But, initially, markets like Gresham's Royal Exchange had to be small clubs. And science, too, originated in a small club.

THE CREATION OF THE FIRST INVISIBLE COLLEGE

One of the most useful books on the sociology and history of science is *Leviathan and the Air Pump*. Written by Steven Shapin and Simon Schaffer of the Universities of California and Cambridge, respectively, and published in 1985, it examines the origins of the Royal Society of London, the world's oldest surviving research society, founded in 1662.

The scientists who created the Royal Society faced novel problems: how did they know when they had made a discovery? What, indeed, was a discovery? And what exactly was an experiment? Those questions seem naïve now, but during the 1660s people were still unsure if the making of discoveries was even possible, let alone legitimate. And the greatest opponent of the idea of science as a useful activity was Francis Bacon's one-time secretary Thomas Hobbes (hence the *Leviathan* of Shapin and Schaffer's book).

Hobbes, whose rigid thinking seems to have misled him on every important matter, believed that the only credible facts were those obtained by logic. He lauded geometry because, once its premises had been established, its deductions were 'indisputable'. But Hobbes recognized that scientific knowledge was largely tacit, so he viewed it as hopelessly subjective, and he dismissed experiments as arbitrary and whimsical and therefore unworthy of a true philosopher. He viewed science as we now view alchemy or astrology – as an invalid activity.

In contradistinction, science's great advocate was Robert Boyle (hence the *Air Pump* of the book title). Boyle, one of the most influential of the founders of the Royal Society, proclaimed that new facts could indeed be established, but – and here he proclaimed a novel philosophy – only as a *collegiate* activity. Boyle conceded that new knowledge was tacit and that experiments (being performed by individuals, who must inevitably be arbitrary and subjective) are indeed arbitrary and subjective. Boyle conceded, therefore, that the evidence of any sole researcher could never be accepted *ex cathedra*. But, he contended, new facts could be added to the canon of knowledge if a number of persons of 'unimpeachable' integrity had 'witnessed' the observations that underpinned them. The early experiments of the Royal Society were always communal.

Note Boyle's legal language: 'unimpeachable', 'witnessed'. Boyle drew on the law for his analogy:

> Though the testimony of a single witness shall not suffice to prove the accused party guilty of murder; yet the testimony of two witnesses, though but of equal credit . . . shall ordinarily suffice to prove a man guilty
>
> *The Sceptical Chymist*, 1661, I, p. 486

Boyle therefore helped launch the scientific revolution as a collective activity, where one person's tacit knowledge could be legitimized on its being tested by others; and this collegiate philosophy was endorsed in 1667 when Thomas Spratt, in his official *History of the Royal Society,* condemned individual 'pride, and the lofty conceit of men's own wisdom'. Truth, he said, could advance only through the collective and 'unanimous advancement of the same works'. The men who created the Royal Society, therefore, consciously created a new philosophy of truth, one obtained by collective experience.

But who was part of the collective? Obviously, only trustworthy people could be admitted, which was why Boyle described his lab as an 'Elysium . . . a seat of bliss' across whose 'threshold' only few people might pass. From the beginning, therefore, the Royal Society excluded non-scientists (other than aristocrats of the rank of baron and above) from its meetings and from its public experiments. In his 1665 *Free Enquiry* Boyle wrote that attendance at the Society's public experiments was not to be extended to: 'even of the learned amongst logicians, orators, lawyers, arithmeticians &c [because] they are not physiologers [experimentalists].' Like access to the markets of medieval Italy or the Royal Exchange, admission to science was restricted.

Hobbes himself was refused admission to Royal Society experiments, so he contended that the Royal Society was not a public place and that its definition of a fact – that it be publicly validated – was false. Its facts were validated only by a mutually-selected club. And Hobbes was right, science was never a public good, it was always only a collegiate one. And it needed to be collegiate because its members had to trust each other to report their findings honestly.

Ah yes, report. Because Boyle's first task was to get his fellow scientists actually to report. As we have seen, scientists want to claim priority in making discoveries, but, to keep their lead, they do not want to tell others how they made their discoveries. But Boyle used his leadership within the Royal Society to negotiate both the convention whereby priority – and therefore esteem – goes to the scientist who publishes first, not to the scientist who might have made the discovery earlier but who has kept the findings secret, and the convention that papers are accepted for publication only if they contain a methods as well as a results section, to allow reproducibility.

We see here, therefore, that science is not innately a public good: it is innately a *discreet* one where, in a state of nature, scientists would publish not their methods but only their findings, and sometimes delay the publication even of those. But it was Boyle who realized, in classic game theory mode, that if members of the Society collaborated with each other in publishing their findings openly and including their methods sections, then all the scientists within the Society would do better. And because the Royal Society's publications were effectively restricted to fellows, the fellows enjoyed an advantage over non-fellows.

Science, therefore, only appears to be public because, over the centuries, most scientists globally have gradually modelled themselves on the Royal Society's 'new' conventions. Yet not all scientists have! Consider the protein crystallographers. It takes so many years to discover how to crystallize any particular protein, that the protein crystallographers have elaborated the convention of publishing their findings a year before they publish their methods, thus keeping a lead on the further study of 'their' protein. Everyone agrees that, since the copying of a protein crystallization method is so much easier than discovering it, this is only fair.

During the late 1990s some general journals such as *Nature* or *Science* abandoned the protein crystallographers' convention, but some specialized journals still observe it. This is a significant survival of an arcane practice because it shows that the now-near-universal publishing of science is *not* because science is inherently a public good but because its publication is a self-interested game-theory activity. And when a particular field agrees collectively that its members' self-interest is not served by free publication, it elaborates different customs.

SCIENCE IS A CLOSED WORLD

To this day a scientific conference is not open to the general public but only to members of a club, literally. The conferences I attend are those of the Society of Investigative Dermatology, the European Society of Dermatological Research and the Biochemical Society, and access is restricted to members of the appropriate societies, whose membership, like the membership of Pall Mall clubs, is restricted to persons of good repute who, as trustworthy scientists, have been nominated, seconded and voted in. And because conferences are actually private trading floors, an interesting phenomenon can be seen: hierarchy. Scientists do not chat willy nilly. The top researchers talk only to each other, average researchers are consigned to the company only of other average researchers, and non-researchers (such as those industrial scientists who do not publish) are ignored.

Science is hierarchical because relatively few scientists matter. Back in 1926 the biomathematician Alfred Lotka had already noted that a mere 6 per cent of researchers account for 50 per cent of all papers and – because quantity and quality go together in science – almost all of the papers that matter. That handful of leaders is interested in trading knowledge only with fellow leaders, because only they have something to trade, so the leaders speak only to each other, ignoring hoi polloi.

So a conference today may appear to be an insane forum where researchers give away information for nothing, but it is, in fact, a trading floor where, in the language of the game theorists, scientists have moved from a Nash equilibrium (where everybody has optimized their own position relative to everybody else) to a Pareto optimum (where science at large has optimized its position to its individuals' greater collective benefit).

Journals only appear to be public

If science is a self-interested activity only masquerading as philanthropic, then journals should not be public goods – and they are not. Consider one of the great hypocrisies of contemporary science, namely peer review. When papers are submitted for publication, they are sent to referees for 'dispassionate' assessment. But the assessment is not dispassionate! When journals, as a naughty experiment, send out copies of the same paper (but with different authors in each case) to different referees, the papers that purport to come from famous scientists are generally accepted, but those that apparently come from unknowns are generally rejected.

Which is how it should be. How can a referee assess a paper without assessing the likelihood that the authors have done and seen what they claimed? Because so much science is tacit and therefore hard to repeat, and because we no longer perform our experiments in public at the Royal Society, fraud is easy, so trust has to be earned. Fraud is indeed common. In a recent survey of several thousand scientists at the National Institutes of Health (a prestigious US institution) 15 per cent of the researchers admitted to changing their methodology or results, or to ignoring inconvenient observations, to please a funder.[25] And that's only what the scientists admitted to! Repeated surveys have concluded that between 1 and 2 per cent of all papers are either 'fabricated, falsified or plagiarized'. Consequently, journals have to work hard to ensure that only scientists with good reputations get published readily. Equally, only reputable scientists get research grants.

Thus we see that science is a social institution based primarily on mechanisms that reinforce trust. This is where the postmodernists such as Feyerabend have got it wrong: science may look like a cartel of powerful individuals supporting each other in a parody of Matthew 25:29 ('to him who hath yet more shall be given'), but in fact it is a college of honourable people struggling against the never-ending problem of fraud. So publication in scientific journals is open only to trusted members of the club.

Equally (and importantly) the readership of journals is also limited, because only fellow scientists can actually understand them. We are told that science is a public good because anyone can access the journals. Oh yeah? How many people can understand Einstein's papers on relativity? Those papers are over a century old, so think how much less penetrable contemporary papers must be.

Consider as an analogy the law. Is that a public good? It appears to be. It's written down, and anyone may access the statutes and the law books. But those statutes and books are impenetrable to non-specialists, and 'the person who represents himself in court has a fool for a lawyer'. It takes years of study to master the subtleties and lore of the courts, so litigants employ lawyers to represent them. To non-lawyers, therefore, there is no disembodied entity called the law, there are only lawyers. Science is like that. In theory the journals are public, but in practice only certain experienced scientists can access them.

So, for example, in his 2007 autobiographical *A Life Decoded* Craig Venter described how, after he had turned down the offer to join the entrepreneur Rick Bourne in sequencing the genome, 'after me, he [Bourne] believed Sulston and Waterson were the best propects for sequencing, and if they left he would have no program' – that is, far from being a public good, major high throughput gene sequencing then was the expertise of precisely two research groups on the globe.

This has economic implications: to entrepreneurs, there is no such thing as science, there are only scientists, the best of whom exercise near-monopoly control over new knowledge. In the formal language of the economists, oligopolistic scientists are excludable and rivalrous – and private goods.

THE ROYAL SOCIETY

The Royal Society, therefore, was created by Boyle and his colleagues as an institution to reinforce trust. And it needed to be based on trust because its creation was remarkable – its first fellows had no reason to trust each other. The ten men who, on 28 November 1660, after listening to a lecture of Christopher Wren's on astronomy, stayed behind at Gresham College in London to found the Royal Society, were disparate indeed. Some, like Boyle, were aristocrats, while others were of modest origins. More dramatically, half were Royalists and half Parliamentarians: in the immediate aftermath of England's bitter Civil War, Commonwealth and Restoration (1642–60) how could these enemies join the same club? How, indeed, could King Charles II promote the Royal Society so enthusiastically? Charles hated Cromwell yet he appointed, as the Society's first president, none other than Sir Robert Moray, Cromwell's brother-in-law! And Boyle's brother, Lord Broghill, was one of Cromwell's close lieutenants. How could this king and these men even meet together, let alone promote science?

Because they were all Freemasons. The 2002 book by Robert Lomas, *The Invisible College*, clarifies the tale. Everyone concerned, even the king (though the evidence in his case has subsequently been obscured) was a Mason. The men were Masons first and scientists second, and they modelled their institution on Masonry's. So the Royal Society's statutes, which forbade religious or political discussions, which imposed institutional rules of egalitarianism, honesty and courtesy, and which restricted admission to the carefully selected, were modelled on Masonic prescriptions. The club came first, the science second.

Today, many people view Masonry with suspicion, but in earlier and less democratic ages the forces of good needed sometimes to work discreetly. In tribute to the Founding Fathers the back of the US $1 dollar bill is decorated with Masonic images, from the pyramid to the Eye, yet most of us would allow the introduction of republican democracy into America to have been a 'good thing'. Gresham, Francis Bacon and King James I were all Masons, and both the Royal Exchange and Gresham College, like the Royal Society, were initially Masonic clubs. The club moralities came first, and the market – like science – followed. Rarely has history illuminated a social truth so vividly: the collective moralities had to precede their fruits of wealth and truth.

And clubs have rules. In his *Sceptical Chymist* Boyle stated that the scientists of the Royal Society needed to be: 'sober and modest men . . . diligent and judicious . . . drudges of industry.' Boyle required them to be courteous and to avoid 'contumelious language'. They could criticize only a man's 'observation, not his want of sincerity'. To reinforce courtesy, Boyle invoked Vespasian's Law, an early version of game theory's tit-for-tat: 'It is uncivil to give ill language first, but civil and lawful to return it.' In denouncing contumelious language, Boyle was addressing a contemporary problem. The intellectuals of the day could be horribly robust. Here is Hobbes attacking John Wallis and Seth Ward, two Oxford dons (in Holy Orders as they all then were) for questioning his maths:

> So go your ways, you Uncivil Ecclesiastics, Inhuman Divines, Dedoctors of morality, Unasinous Colleagues, Egregious pair of Issachars [crooked sophists], most wretched Vindices [dishonest pseudonymous advocates] and Indices Academiatrum [bloated reactionaries].
>
> *De Corpore*, 1655

We can see why Boyle excluded Hobbes from the Royal Society. Not only was Hobbes no scientist, so he possessed no useful knowledge to trade, but he was not clubbable either.

SCIENCE IS AN INVISIBLE COLLEGE GOOD

So, what is science? Let me answer that by following the career of a young scientist, assuming for simplicity that he is male. He is born, as we know, with an innate creativity, an innate understanding of the scientific method and an innate ambition for esteem, but is otherwise ignorant. At around eighteen he leaves (high) school knowledgeable, but only of public knowledge, namely the old science that is available in the text books. At around twenty-two, he graduates from university, also knowing largely only public knowledge (though the better universities expose their students to some 'insider' knowledge, and to some of the working practices of effective researchers).

Thereafter, he enters into an apprenticeship that is almost medieval in its progression. The PhD is a key rite of passage, and studies on the sociology of science confirm that almost all effective researchers will have undertaken their PhD in the lab of one of the 6 per cent of scientists who matter (who will almost certainly work in one of the handful of elite universities or research institutions).[26] Equally, if our young scientist is to flourish, his three to five years' subsequent post-doctoral experience must be undertaken in the labs of another one or two of the 6 per cent. Only after our young scientist has negotiated these steps successfully and has co-authored good papers in consequence, will he be accepted by the community as a peer with whom conversations at conferences can usefully be held and who would be eligible for a tenure-track job from which to research freely and from which to apply for research grants.

Why is the apprenticeship so prolonged, and why is the quality of lab supervisor so important? Because of tacit knowledge and of its accompaniment, the centrality of trust. Only someone who has worked alongside major researchers will acquire (painfully) all the tacit knowledge of how to do science, in all its myriad aspects, and only someone who has successfully passed the moral scrutiny of major researchers will be trusted by the rest of the community. Which tells us what science is. Science is not the observations that an individual makes, because those observations have no traction unless others accept them. Thus the flawed observations of frauds or fools will not be trusted or accepted.

But there are drawbacks to science being collective: the work of genius can equally be rejected. Gregor Mendel (of mendelian genetics) published his great paper 'Experiments on Plant Hybridization' in 1866 in *The Proceedings of the Natural History Society of Brunn*, but it was not 'science' because no one understood it. Only after 1900, when Mendel's paper was rediscovered by a scientific community that had finally understood the question Mendel was answering, did it become 'science'.

Equally, Alfred Wegener's 1912 theory of continental drift was not 'science' for over forty years; only when it was discovered that the magnetic fields of rocks of different ages aligned differently did the idea gain support and become 'science'. On the other hand, the idea of eugenics, which we now reject, was 'science' for forty-five years between the 1900 rediscovery of Mendel's work and the fall of the Third Reich in 1945.

Science, therefore, is the collective view of a group of people that their published work is good. In the postmodernist cliché, 'truth is what we say it is'. And who are 'we'? 'We' are organized in invisible colleges.

SCIENCE IS ORGANIZED IN INVISIBLE COLLEGES

The Royal Society originally met (for fear of the Church) secretly as the 'Invisible College', tucked away within Gresham College, London, and Wadham College, Oxford. Only after the Civil War could its members come out. But, as the physicist Derek da Sola Price noted, science is still organized in 'invisible colleges'.[27] Originally there was only the one invisible college, but as science has grown and sub-divided into separate fields of research so the scientists in each field have spawned their own invisible colleges.

I am a member of the invisible college of experimental dermatology. I publish primarily in the *Journal of Investigative Dermatology* and the *British Journal of Dermatology* and I speak generally at these journals' society's meetings. I also interface with other invisible colleges, particularly in medicine and cell biology, but I don't 'feel' a full member of them.

But science grows inexorably, and as each invisible college expands so new sub-colleges are born. I increasingly attend the meetings of the new European Hair Research Society and the even newer European Sebaceous Club. Eventually, these will each spawn their own invisible colleges. New invisible colleges are born, in part, because of the sheer size of science: no single person can grasp more than a tiny part of contemporary research; introduce me to inorganic chemists or plasma physicists and I am lost.

But an invisible college is more than a field of science: an invisible college is primarily the human forum where researchers meet in conferences and where they review, publish and read each others' papers. New invisible colleges, therefore, spin off when the old ones become too large in terms of the number of people. Why? Because only when a group of people is small enough to know each other by reputation can they publish each others' papers, attend each others' talks, trade information and support each others' grant applications with confidence.

Because knowledge is so tacit and thus so easy to misrepresent, the world of science is essentially a trust-maximizing institution. It is therefore essentially a group of people who have learned to trust the reports of others.

* * *

Science is a conversation held between researchers who have learned to trust each other and who share similar tacit experiences. Free-riding is not a problem, therefore, because entry to the invisible college is earned only by frequent high-grade publications, so scientists' apparently 'free' access to others' science is actually paid for in advance.

The economists have yet to model science satisfactorily. The best models we have for science come from the historians and sociologists. So it was Robert Merton, a sociologist, who in his 1942 essay 'The Normative Structure of Science' described science with the acronym CUDOS (note how it is pronounced). The letters stand for Communism, Universalism, Disinterestedness and Organized Scepticism, by which Merton meant that scientists share knowledge (communism), that knowledge is judged objectively (universalism), that scientists act in ways that appear selfless and that ideas are tested collectively.

No economist has come close to modelling CUDOS or the invisible college mathematically, but economics is an important discipline, so its failure to model science is damaging. The economists possess a number of standard models (public good, private good, merit good, club good), but none describes science. So, though science has features of both a private good (companies invest in pure science because it is profitable) and of a public good (most of the benefits of a company's research are snapped up by other companies), science obviously is neither. Nor is it a 'merit good'. 'Merit' goods are goods that the market will supply, but only suboptimally (thus requiring government supplementation), but because the government funding of science provides no economic boost (see later) science cannot be a merit good.

The economists describe a class of hybrid good, the club good, which might describe science. Consider a tennis club. That seems to have both private and public aspects. A tennis club is certainly a private good, and people have to pay to join the exclusive group of lithe, bronzed athletes who besport themselves on the courts. But, for actual members of the club, the facilities are 'public' (as defined by the economists, namely that members may use the courts at no extra cost, so any game is free to them, and there are lots of courts, so one member's use of a court does not stop another member using another court; in technical terms non-rivalrous and non-excludable). The tennis club is, therefore, neither a private nor a public good but a hybrid, which the economists call a club good.

Science is a bit like that. To Jack of Glaxo, listening to Black's talk, the science was freely available. To the ten Japanese pharmaceutical companies of Odagiri and Murakimi's survey each other's science was freely available. To scientists of the same invisible college, therefore, each others' science is freely available. But to access the science 'freely', scientists and their employers have to pay a huge membership fee, namely the long years of apprenticeship and the on-going publishing of papers.

But clubs compete with each other (there is more than one tennis club out there, competing for new members) whereas invisible colleges do not. So we cannot piggyback on the economists' experience of club goods, we have to describe science as a unique good, one that reflects what we actually know it to be: science is an invisible college good.

What does that mean? It means that any particular area of science is understood by only a few cognoscenti, who trade knowledge for mutual benefit. And the trade is unusual because it is not a simple barter of A for B between two individuals, but, rather, it is more like the pooling of information between peers. Any particular discovery may benefit others more than the discoverer, yet over a period of time, with enough pieces of information being pooled, chance will ensure that the advantages are distributed between all players.

And the idea, held by the economists, that this elaborate dance is some sort of public good, is misleading.

15

The Niche is Everything in
Biology and the Markets

'An economist is someone who, on being shown something that
works in practice, wonders if it would work in theory.'

Ronald Reagan, 1987

In March 1981, following Margaret Thatcher's government's budget of
that year, the leading 364 economists of Britain (including the Nobel
laureates) wrote a letter to *The Times* newspaper condemning the
budget as having 'no basis in economic theory' and as being likely to
cause such economic collapse as to threaten Britain's 'social and
political stability'. During the budget debate in the House of
Commons, Mrs Thatcher was asked if she could name even two
economists who supported her policies. She named Patrick Minford
and Alan Walters, but, it is said, she was relieved not to have been
asked to name a third. Yet, as everybody knows, the recovery of the
UK's economy following the long post-war decline can be dated to the
budget of March 1981.

It is easy to mock economists, and the modelling of something as
complex as a modern economy is not easy, but on the subject of
science the economists have been as unhelpful as they were over
Britain's budget in 1981. The problem can be summarized by the
quote from the economist Paul Romer:

Research has positive external effects. It raises the productivity of all future individuals who do research, but because this benefit is non-excludable, it is not reflected at all in the market price.[1]

Yes, research does have positive external effects, and they are indeed non-excludable (the benefits cannot all be retained by the original researcher because others can pick up on their ideas), but so what? Why should the benefits to others be reflected in the market price? It is only because of their weird theories that economists can argue that the researchers' market price is too low.

A BRIEF HISTORY OF CLASSICAL AND NEOCLASSICAL ECONOMICS

The so-called 'classical economists', people like Adam Smith, were essentially wordsmiths. They used simple statistics and simple maths, but essentially they were prose writers. 'Neoclassical economics' starts just over a century ago with modern price theory.

Price theory may seem boring, but what *does* determine the price of an object? The story opens with Thomas Aquinas, the thirteenth-century saint and philosopher, who proposed a labour value of price: he suggested that the price of an object was the cost of the labour that went into it, and that if two different objects each took an hour to make, each would cost the same. But, as Hobbes noted in *Leviathan*, the labour theory is wrong: 'not the seller but the buyer determines the price.' Two different objects may each have taken an hour to make, but they may not be equally attractive to purchasers.

It was a group of nineteenth-century economists, including Léon Walras, Alfred Marshall, William Jevons and Carl Menger, who proposed how prices might be determined – by a 'marginalist' approach. They proposed that we live in a world dominated by diminishing returns, both to consumers and producers.

Diminishing returns to consumers

Imagine a market in bread, and imagine you are a consumer. You love bread but you have none. To buy one loaf, therefore, you would pay a great deal of money. For a second, quite a lot of money. For a third, some money. But you'd take a rain check on the ninth loaf because, by then, you'd be full. The price you'd pay for each successive loaf, therefore, falls.

Diminishing returns to producers

Imagine you are a producer and that you own an oven. To produce the first ton of bread you employ a baker. To produce the second ton, you employ a second baker. But to produce a third ton you find you have to employ two more bakers, a third and a fourth: that is because, with each successive hiring, total *production* will rise but *productivity* will fall, because each successive baker increasingly gets in the way of the others and the oven's maximum capacity is increasingly approached. Eventually, a point is reached where employing a further baker would cost more than can be recouped, that cost being (here comes the price) that which the consumer would pay for their eighth loaf.

The price of bread, therefore, is reached when the price the consumer will pay for the final (or marginal) loaf equals the cost to the producer of the final (or marginal) worker. The 'marginalist' revolution, therefore, launched economics as a mathematical discipline, because the economists could use their model of diminishing returns to predict prices. Economics had become a real science, just like physics!

But economics remains an odd discipline because it is driven by theory, not by observations: economists like to generate theories and see if they work in reality, whereas most scientists prefer to observe the real world and then generate their theories. It is this love of theory that makes economists so dangerous. After all, how credible really are the stories of diminishing returns to producers and consumers? An awful lot of reality has to be overlooked to generate such simple stories of how prices are determined. Yet as Keynes noted: 'The world is ruled by little else [but] the ideas of economists.'

And their most dangerous idea is that of the 'perfect market'. In the words of the current edition of the *New Palgrave Dictionary of Economics*: 'No set of ideas is so widely and successfully used by economists as is the logic of perfectly competitive markets. Correspondingly, all other market models are little more than fringe competitors.' But the perfect market is absurd and misleading.

DIMINISHING RETURNS LEADS TO THE SO-CALLED PERFECT MARKET

One major author of 'perfect markets' was the Italian economist Vilfredo Pareto. Pareto (1848–1923) was not always polite (after touring England in 1880 he wrote: 'We are given to believe that the English are a hard-working people – it is but an illusion'), and he was not always realistic either because he claimed he could model an entire economy if he extended diminishing returns across the whole economy.

To perform that mathematical exercise, Pareto wished away all monopolies. Consider bread. In the real world there are hundreds of different types of bread and of bakeries, ranging from Poilâne, the queen of bakeries, via butter croissants to supermarket sliced white. There are as many different customers' preferences. Indeed, with 6 billion potential customers on the planet, thousands of different products and hundreds of thousands of different outlets, modelling the bread trade globally is clearly impossible. But if the economists assume that all loaves are the same, that they all cost the same, and that everyone likes the same bread, then the global bread industry can be modelled. The model would be a fiction, because it would overlook the key fact about the trade, namely that it is complex, but the maths would be neat.

That's what Pareto did for the whole economy. He assumed that no one had more economic power than anyone else. So he assumed that no one knew more than anyone else (so-called 'perfect knowledge'), and he assumed that all producers, all consumers and all products were the same. Under those circumstances, he found in his 'theorems of welfare economics' that his perfect market was not only mathematically but also morally perfect because, under it, the price mechanism would optimize the distribution and production of goods.

It is the beauty of this maths that enraptures the economists, because they can calculate how goods are distributed in their imaginary world. But the perfect market is, of course, a fiction for at least three reasons. First, perfect knowledge is a nonsense. Scientists will always know more science than do, say, musicians. Second, the way Pareto pretended that all producers, all consumers and all products were the same was by assuming that there was an infinity of them. Under the laws of mathematics, an infinity means, paradoxically, that instead of having to worry about an awful lot of entities, you can assume they are all the same, so there might be only one producer, one consumer and one good in the perfect market.

But infinities are mythical. Imagine the market in bread and cakes. As we know, bread and cakes come in hundreds of different flavours, shapes and sizes, but according to the perfect market every bakery you entered – and there would be untold millions of bakeries on your very doorstep – would carry millions of every loaf and cake. There would be millions of apple tarts 6 inches wide, millions 7 inches wide, millions with raspberries too, and on and on. In reality, of course, bakeries are limited by simple physics – the average street can accommodate only a few – and by the numbers of customers. There is not an infinity of consumers, only the handful that live within easy distance of the high street.

Finally, in a 'perfect market' no one makes a profit! In a perfect market the price of goods in the shops would be cost price because an infinity of producers, all competing with each other with perfect knowledge, would drive each others' prices down to cost price: there are no real profits in a perfect market, only so-called 'normal' or technical ones because, with an infinity of producers, interchangeable goods and consumers, the producers cannot be entrepreneurs because they cannot offer any different products from anyone else, nor can they pay themselves more than than they pay anyone else.

The French economist Léon Walras admitted that a perfect market was a fiction: markets in reality are driven by entrepreneurs, but there are none of those in a perfect market: 'Assuming zero-profit equilibrium, we remove entrepreneurs from the economy . . . They make their living not as entrepreneurs but as land-owners, labourers or as managers of their own businesses.'[2]

The real world, of course, is actually constructed on monopolies and oligopolies, which are the very basis of a market economy. In the real world all profit derives from the fact that every producer enjoys some sort of monopoloy or oligopoly power: he is the only baker on the high street or she is the only Mme Poilâne. Obviously no baker can charge prices that are too high, or customers will be driven away, but profits come from oligopoly. Yet such common sense is excluded from so-called 'perfect markets'.

WHY THE PERFECT MARKET HAS DONE SO MUCH DAMAGE

OK, so we've shown that the perfect market is a nonsense, but why get bothered? After all, that section was a bit abstract and this section is going to be even less good reading. But the quote from Romer at the beginning of this chapter is important:

> Research has positive external effects. It raises the productivity of all future individuals who do research, but because this benefit is non-excludable [the benefits cannot all be retained by the original researcher], it is not reflected at all in the market price.

People therefore believe that Romer has proved that government must fund science to compensate researchers for the fact that their research raises the productivity of all future individuals who do research, but that it is not reflected at all in the market price. But Romer has not proved that! Romer has confused material property rights in a perfect market with intellectual property rights in a real one, and never the twain shall meet.

Property rights in things – material property rights – matter in all markets. In a perfect market material property rights have to be respected or the producer goes bankrupt. In a perfect market a producer who loses some of his revenue (to thieves, for example) must go bust because there is no fat to trim – with an infinite number of competitors, every factor of a firm's production must already be maximized at the margin. Something similar is true of property rights in a real market: in the real world people will not invest if thieves steal their property.

Consequently, the economists believe they have successfully extrapolated the importance of material property rights from perfect markets to real ones. True, they have. But not everything can be extrapolated between markets.

Consider the person who builds a bakery in both markets. In a perfect market, the bakery owner can predict to the penny what their return will be, but in the real world a person can build a bakery yet an insufficient number of customers may turn up.

To then extrapolate to say that, because in a *perfect* market a producer cannot fail, therefore in the real world an entrepreneur should be *compensated* by government for failure, is an obvious nonsense. The producer in a perfect market can predict exactly what his revenue should be, but the entrepreneur in a real market is not aiming for revenue, he is aiming for real profits, and in real markets there is no calculation of what someone's profits 'should' be. In a real market, profit emerges out of monopoly, and that is achieved not by charging cost price but by charging consumers what they will bear, which is unpredictable. There is no 'theorem of welfare economics' in real markets that says that entrepreneurs should monopolize their markets indefinitely – rather the opposite in fact, for by what right should an entrepreneur enjoy monopoly prices indefinitely?

So, the argument that a producer will underinvest when his *material* property rights are violated in a *perfect* market is simply not extrapolatable when the intellectual property of a monopolist is violated in a real market. The economists pretend that the criminal theft of the material property rights of a producer in a perfect market is analogous to knowledge transfer in a real market – but it isn't. It's as if the economists are treating apples and screwdrivers as interchangeable. They are not. Apples and screwdrivers occupy different categories of existence.

And Romer was therefore wrong to extrapolate from material property rights in perfect markets to intellectual property rights in real ones.

RESEARCH IN THE REAL WORLD

Let's get away from abstract discussion and see how economists react to research in the real world. Imagine again a baker. To make a profit in the real world, a baker needs a monopoly. Perhaps he (let's pretend it's a he) could invent a novel product – sliced bread, perhaps. To invent sliced bread, the baker takes out a bank loan, he invests in research, he develops a mechanical slicer, and soon he will be selling his sliced bread at vast profit because he's the only producer.

What will his competitors do? Some, stunned by the fall in their revenues and by his profits, will take up their own research. They will take out bank loans and employ researchers. Soon someone will invent thin sliced bread (for sandwiches). That someone will then steal much of the first baker's profits.

At this point a strange phenomenon will develop: the economists will cry 'unfair'! 'Poor first baker,' they'll wail. 'Someone has stolen his intellectual property. People developed thin sliced only because of the first baker's idea.' Which is true, but thereafter the economists go awry. You and I know what the first baker will next do. He'll revisit his bank, take out another loan, research even more intensively than before, and develop thick sliced for toasting. Soon he will be enjoying huge monopoly profits as people buy his bread to make toast. You and I, therefore, will argue that the various bakers will create the best possible world for customers, perennially improving their products by their competitive research.

But the economists argue the opposite. They say that, because someone invented thin sliced, the first baker's profits from the initial invention of slicing are lower than they would otherwise have been. Therefore the first baker will research less than he would otherwise have done. Moreover, the first baker will downsize even further because he can anticipate his diminution at someone's hands: even if the first baker invents thick sliced, he knows that someone will soon invent muffins or some other market-stealer. And everyone else will do the same. They, too, will anticipate that others will acquire their ideas, so they, too, will research less.

Therefore, say the economists, competition will cause entrepreneurs to reduce their research budgets. The economy will thus stagnate, and only the government can restore the nation's research budget to the optimal level. Only if a company holds an unshakeable monopoly, the economists say, will it invest properly in research. But the economists are transparently wrong, not only because they also teach that monopolists need do no research, and not only because they have extrapolated illegitimately from perfect markets to real ones, but also because the empirical evidence shows clearly that . . .

COMPETITION SPURS RESEARCH

In reality, competition is the great spur of research. In a recent survey of 154 Spanish research-led companies the economist Isabel Busom found that the majority confirmed that they 'would accelerate their own R&D effort if they found that a rival firm was doing similar R&D'.[3] Similarly, in a survey of agricultural R&D across the developing world, Carl Pray and Keith Fuglie of the US Department of Agriculture found that even in poor countries, it is market competition that spurs private research:

The most liberal market economies of the 1980s – Thailand, Malaysia and the Philippines – had the highest private research intensities. The countries with the most controlled economies – China, Indonesia, Pakistan and India – had the lowest. The countries in which private

346

research grew most rapidly – China, India, Pakistan and Indonesia – had major liberalization programs during the mid-1980s.

US Department of Agriculture, *Agricultural Economic Report*
No.805, 2001

A disgraceful episode of US history, namely the War of 1812 (the Second Anglo-American War), confirms that competition incentivizes research. In 1812 the Americans, allied to the tyrannical Napoleon, attacked Britain, the world's sole defender of freedom. But in 1814 the Americans, from their Blackened House in Washington, D.C., were forced to sue for peace.

It was the economic historian Kenneth Sokoloff who used the peace following the War of 1812 to test the effects of trade on the USA. The war disrupted trade between British Canada and the USA for up to three years, but thereafter trade restarted. What were the consequences on the northern US counties adjoining Canada? Sokoloff found that the resumption of trade prompted research. With each new incursion of trade after 1814, local businessmen started to patent.[4] The sudden rises in local patenting, moreover, were not caused by influxes of inventors from Boston, New York and other areas of existing innovation; they were caused by local businessmen turning to research to defend their existing businesses against new competitors and to exploit new markets to which the new trade routes had provided them with access. Competition stimulates research. As the head of R&D at Unilever used to say, his department's budget's best friend was the R&D department at Procter & Gamble.

We thus see that economic growth is driven by monopolistic (or, technically, oligopolistic) competition, and that companies invest in research when faced with competitors. One contemporary economist who understands this is William Baumol of Princeton, who wrote as the very first sentence of his 2002 book *The Free-Market Innovation*

Machine: 'Under capitalism, innovative activity – which in other types of economy is fortuitous and optional – becomes mandatory, a life-and-death matter for the firm.' This is because, as Joseph Schumpeter wrote in his 1942 book *Capitalism, Socialism, and Democracy*, economic theories based on perfect markets and the price mechanism are unreal – in real life companies compete for monopolies by innovation:

> In capitalistic reality as distinguished from its textbook picture, it is not that kind of [price] competition that counts but the competition from the new commodity, the new technology . . . which commands a decisive cost or quality advantage and which strikes not at the margins of the profits and the outputs of existing firms but at their foundations and their very lives.
>
> J. Schumpeter, 1947 (2nd edition), p.84

Those pressures have led to research having been routinized:

> Innovation itself is being reduced to routine. Technological progress is increasingly becoming the work of trained specialists who turn out what is required to make it work in predictable ways.
>
> J. Schumpeter, *Capitalism, Socialism, and Democracy*, 1942, p.132

And the economist of technology Jacob Schmookler agreed:

> Invention was once . . . a nonroutine economic activity, though an economic activity nonetheless. Increasingly, it has become a full-time, continuing activity of business enterprise, with a routine of its own.
>
> J. Schmookler, *Invention and Economic Growth*, 1966, p.208

We see here, therefore, the destruction of the theories of perfect markets. The theories, with their associated 'theorems of welfare economics', are intellectually attractive but irrelevant to real life. Thus have the economists spun stories of 'market failure' in science, which is why even a good economist like Richard R. Nelson can lament:

> It is something of a puzzle, therefore, why the capitalist innovation system has performed so well. There is certainly nothing like the twin theorums of welfare economics around to support an argument that 'capitalism can't be beat'.
>
> R. Nelson, *The Sources of Economic Growth*, 1996, pp.54-55,.

But monopolistic competition *does* show how capitalism can't be beat. Because everyday trades are uncertain, it cannot show it with the mathematical precision of perfect market theory – just as Heisenberg's Uncertainty Principle demonstrates that we cannot observe the physical world, either, with absolute certainty – but entrepreneurs cannot be the profit-maximizers of conventional economic theory, they can be only profit-seekers, and some entrepreneurs will do 'better' than they 'deserve' and others worse, but those are chance outcomes that do not undermine the fundamental truth that, when returns increase (as they do with research), monopolistic niche-based competition must be the outcome, and a healthy one it is too.

A FIRM'S MONOPOLY IS A SPECIES'S NICHE

One of the best models for understanding economic growth by markets was provided by Charles Darwin's theory of evolution by natural selection, and the 'niche' proves the model. Just as there is no perfect market in commercial reality, so is there no perfect market in

Nature. Nature may be red in tooth and claw, but it is not composed of an infinity of competitors. Actually, surprisingly few species compete with each other. The rabbit and the deer might compete for grass, but no polar bear will. Nor will eagles or octopuses. In Nature each species occupies a *niche*, one to which it – and it alone – is uniquely adapted. Nature is a monopolist, and each niche is occupied by only one species. Similarly, in business, companies occupy niches and they compete with relatively few competitors. A PC cannot do duty for a carrot, and it doesn't have thousands of competitors, it competes with Macs.

An early thinker who noticed that niches were integral to business was the American sociologist Thorstein Veblen (1857–1929). In Veblen's time, as today, economics was dominated by concepts of diminishing returns and perfect markets, but Veblen, a sociologist who observed the world as it actually was, saw increasing returns. First he saw it for consumers and in his 1899 book *The Theory of the Leisure Class* showed that consumers engage in conspicuous consumption. Under the theory of diminishing returns, a second pair of shoes is less valuable to a consumer than the first, but to Imelda Marcos her 1,201st pair of shoes was the more pleasurable precisely because she already owned 1,200 pairs

And in *The Theory of Business Enterprise* (1904) Veblen showed that production, too, is dominated by increasing returns: once a company has established its dominance of a niche, it can re-invest its monopoly-derived profits in research and marketing to sustain its dominance. Veblen was not the first to report this; in *Das Kapital* (1867) Karl Marx had noted how markets are dominated by small numbers of large oligopolistic firms that reinvest their profits to further increase their returns. Marx noted that they exploited their economies of scale:

The battle of competition is fought by . . . the productiveness of labour, and this [depends] on the scale of production. Therefore the larger capitals beat the smaller.

And the larger capital beats the smaller because the larger capital can afford to invest more of its profits in research:

The bourgeoisie cannot exist without constantly revolutionizing the instruments of production.

K. Marx and F. Engels, *The Communist Manifesto*, 1847

And in 1912 Schumpeter (the man who coined the term 'creative destruction' to describe the process whereby new technologies displaced old ones) wrote his *Theory of Economic Development* to show that every firm in the real market should make profits by monopoly, because it is by profit that firms find the money to invest in innovation.

It was Alfred Chandler of the Harvard Business School who confirmed that, in practice, firms occupy profitable niches by virtue of increasing returns. In his 1990 *Scale and Scope: The Dynamics of Industrial Capitalism* Chandler systematically recorded the histories of the 200 largest manufacturing companies in the US, UK and Germany of the previous 100 years, and he found that, contrary to myths of perfect competition, the firms were not subject to their markets, they *dominated* them. They did not take prices, they *set* them. They were the firms that had created the technologies and management techniques that had transformed their markets, and, by sustaining those initial advantages by continued innovation, they had retained their domination of their markets and kept out competitors. Thus does business, like Nature, operate by niche monopoly.

It is because monopolies defend themselves by reinvesting in research that they benefit society. Bill Gates's Microsoft has long enjoyed a near monopoly, but it has researched hard to protect it ($4.4 billion annual research budget). A detailed analysis of manufacturing firms has confirmed that the higher the market share of a company, the greater its investment in R&D.[5] As the Italian economist Federico Etro showed in his paper 'Innovation by Leaders', we thus see a virtuous cycle in which research has created growth that has funded further research to maintain that company's market share.[6] If firms fail to protect their monopolies by reinvesting their profits into research and the other sources of monopoly, their competitors will displace them.

Equally, species in Nature have to innovate constantly to protect their niches from competition. Obviously species do not innovate by research, but they do have sibling variation, and it is the variation provided by each new generation that provides the innovation a species needs to defend its niche and to adapt if its niche changes. For sibling variation to yield innovation there must be sibling excess, but being a monopolist the species creates 'profits' in the form of an excess biomass, which it invests in lots of progeny. The analogy with firms investing their profits in R&D is good.

And because species evolve in response to competitors, the greatest biodiversity is seen in the tropics, because the more fertile the environment the greater is the biomass. The greater the biomass, the greater is the competition, so the greater is the product differentiation or number of different niche-seeking species. This may seem obvious, but it is the opposite of what a perfect market would propose: the economists invented the perfect market to create an infinity of products so they could treat them as interchangeable – that is, they could pretend there was only one product on the market – but in reality, the greater the biomass the greater the number of species, so the greater the variation.

And it is the competition for niches that creates efficiency. So the tree kangaroos of northern Australia and New Guinea can be stunningly (indeed, hilariously) clumsy because, in the absence of competitors for their niche, they have not needed to be more dextrous. There is not some optimal level of dextrousness in Nature that some God-like figure allocates according to a theorem of welfare evolution; dexterity evolves in response to competition. Equally, large markets are more efficient than thin ones, not because large markets approximate to perfect ones in creating interchangeable products but for exactly opposite reasons: large markets create a greater diversity of goods and companies and consumers.

Indeed, Nature is very 'wasteful'. The species whose males fight over females, or whose males eat the offspring of other males, or which produces more offspring than will survive, is not 'efficient' or optimizing 'welfare' in the Pareto sense; just as 'creative destruction' (the supplanting of good but old technology by better, new technology) is also 'wasteful'. But those are the optimal strategies both for evolution and for economic growth via niches.

Spandrels

In Nature species dominate their niches, which resolves, incidentally, a biological conundrum. Stephen Jay Gould argued in his theory of punctuated equilibrium that the evolution of species is sudden, and that most species change very little once formed: 'Evolution does not require gradual change ... sudden appearance and stasis dominate the fossil record.'[7] If species, like firms, dominate their niche, then species will indeed survive unchanged, even in the face of significant environmental or niche alterations, because their increasing biological returns will sustain even suboptimal adaptations. Thus not every biological expression needs to be fully adaptive: Gould's spandrels may be real.

POST-NEOCLASSICAL ENDOGENOUS GROWTH
THEORY

It has taken a long time for mathematical economists to catch up with the observations made by people like Schumpeter and Schmookler. As we know, both Francis Bacon and Adam Smith said, correctly, that scientific research creates economic growth by increasing returns, but until recently the economists of science could not handle that concept. This drove them into absurdities. Consider 'learning by doing'.

Because the economists of science believed that science was a public good, they maintained that industrialists would not fund it. But of course industrialists do. That gap between theory and practice embarrassed the American economist Kenneth Arrow, so in 1962 he proposed that since economic theory can account only for an industrialist investing in a private good such as capital or labour, why don't the economists pretend that science is a type of labour?

As a metaphor, Arrow invoked 'learning by doing'. During the Second World War, the shipyards in America that built the new Liberty ships increased their productivity over time, even though the yards did not invest in science. The men who were employed as labour 'learned' to raise their productivity simply by 'doing', so their increased productivity was an unaccountable consequence of their employment as labour.

Equally, Arrow proposed that economists should model not science but scientists, modelling scientists as labour who improved productivity unaccountably, simply by working as scientists. Consequently the pundits, instead of admitting that they could not model the economics of science, pontificated on 'learning by doing' as if they had described an important phenomenon rather than a metaphor.

Actually, because we now know that science is an invisible college good, and that much of it is tacit and held within the brains of scientists, we can see that Arrow had hit on a good idea – science could indeed be modelled by economists as a form of private human capital. Nonetheless, the economists needed a better model than Arrow's, because they had yet to show mathematically how science might create economic growth. Only in 1990 did the economists find their hero, a University of Chicago professor called Paul Romer.

Romer assumed, following Bacon, that science was a series of discrete bits of public knowledge, but he also noted that bits of new knowledge were, temporarily, partly private goods, because scientists can keep their discoveries temporarily secret (and because they can also patent them). Consequently, new pieces of science can be monopolized and can thus be treated, in part, as private goods. To capture this idea, Romer created a mathematical model where researchers access previously published science freely before selling their newly produced research to monopoly purchasers who then sell the technology on to capital goods producers.[8]

The model is largely fictitious, as are most economists' mathematical models, but nonetheless Romer had apparently modelled how the market could fund science. His theory is now famous, therefore, as endogenous growth theory (because the market itself 'endogenously' funds science rather than depending on science being funded by an 'exogenous' source such as government or God), and the economists now shove aside learning-by-doing to visit the White House or the TV studios and boast of their clever theories. But endogenous growth theory has told us nothing about the real world; it's merely filled a professional gap for mathematical modellers.

Nonetheless, it's important because it filled a key gap. Under Romer's model, firms use science to produce increasing, not decreasing, returns to scale, and they produce – wait for it – profits! Under perfect markets there was no economic growth: there can be no competitive advantage, no money for R&D, and therefore no growth in a perfect market in theoretical equilibrium; there can be only an efficient allocation of existing resources. Perfect markets, therefore, are only a fantasy, where standards of living do not rise. But thanks to Romer, the economists have moved beyond neoclassical economics to post-neoclassical economics, where economic growth and profits can be modelled mathematically.

THE ECONOMICS OF SCIENCE

But the economists are still in a muddle over science, and, for all his virtues, Romer is partly responsible for that. The economists say two contradictory things. On the one hand they say that the flow of ideas is bad for entrepreneurs, yet the flow of ideas is good for society.

The flow of ideas is bad for entrepreneurs

> Research has positive external effects. It raises the productivity of all future individuals who do research, but because this benefit is non-excludable, it is not reflected at all in the market price.

Consequently, Romer says, the market will under-fund science because its benefits, not being excludable, are not reflected in the market price.

The flow of ideas is good for society

Yet if private monopolies in ideas *could* be sustained, and if researchers could keep their ideas secret, the damage to society would be grievous, because most people would be reduced to obsolete technologies and tools. This idea was stated by Thomas Jefferson, among others:

He who receives an idea from me receives it without lessening me, as he who lights his candle at mine receives light without darkening me[9].

Ideas are like a candle: when I apply the laws of thermodynamics, my use of those laws neither exhausts them nor diminishes anyone else's use of them. Therefore, society most benefits when inexhaustible intellectual constructs spread freely, allowing everyone to use them as truly public and truly good.

These two beliefs are, of course, opposed. If we help entrepreneurs monopolize their research (by allowing them intellectual property rights such as patents) they will, it is claimed, do more research, but society will be harmed because it will be deprived of those ideas. But if we encourage the free use of new ideas, entrepreneurs will not fund their creation. A conundrum!

The conundrum, the readers of this book will appreciate, is not real, at least not for science. Literary texts, perhaps, may behave as public goods, because many people can access them, and they can, moreover, copy them cheaply (so perhaps authors need copyright), but science is not easily accessible, nor is it easily copied; it is an invisible college good whose benefits are restricted to members of the invisible college. The person who aspires to light a research candle at Jefferson's must first pay to join the invisible college by constructing their own research candle, whose construction will shed light on other research candlers. There is, therefore, no conundrum in science.

But different governments have, over the centuries, treated the conundrum as real, and they have handled it in different ways. Today, governments subsidize the production of new science, but Jefferson's solution was different: he proposed the abolition of intellectual property rights! Jefferson did not want to abolish all property rights. He advocated property rights in tangibles, including, of course, black people (as Sally Hemings and his 186 other slaves discovered), but he was opposed to intellectual property. When in office Jefferson did, admittedly, introduce a patent law, but only to encourage publication, not monopoly.

Today, we believe the opposite from Jefferson. We do not believe in property rights in people yet we all now worship intellectual property. The classic statement was made by Douglass North, the economic historian:

> The failure to develop systematic property rights in innovation up until fairly modern times was a major source of the slow pace of technological change.
>
> D. North, *Structure and Change in Economic History*, 1981, p.164

Yet it was Jefferson, not North, who was correct. Intellectual property rights (IPR) are a menace, as we shall see later.

AN UNCERTAINTY PRINCIPLE

A visit to the Economics Faculty Library at Cambridge University reveals an odd phenomenon: go to the *Journal of Political Economy* for 1950, look up the paper on pages 211–22 in volume 58, and you will find it torn, ragged and barely legible thanks to all the greasy hands that have pawed it. Even more remarkably, the actual paper itself is not the original; the librarians have rebound the volume more than once over the last fifty years, each time replacing the worn-out pages 211–22 with photocopies. This remarkably studied paper is entitled 'Uncertainty, Evolution and Economic Theory', and it was written by the American economist Armen Alchian who, happily, is still alive. And the paper says that economic growth can be understood only as an evolutionary phenomenon.

Before Alchian wrote, people had believed that economic growth could be planned rationally, not only by governments but also by companies. Companies would assess opportunities and invest their resources accordingly: Mice Catching Mega Corporation might decide that the world needed a better mousetrap, it would assign a budget to its R&D department, the researchers would produce a better mousetrap and, hey presto, Mice Catching Mega would be mega indeed.

Except, as Alchian pointed out, there are no 'hey prestos' in research. The Mega researchers might, quite simply, fail to produce a better mousetrap, or in developing a stronger spring they might, inadvertently, invent a better wire, and the company would diversify and call itself Wired instead. So, for example, when Upjohn was trying to develop a new treatment for high blood pressure, they discovered instead a therapy for baldness (Regaine, Rogaine, minoxidil). When Pfizer was trying to develop its own new treatment for high blood pressure, it discovered instead a therapy for impotence (Viagra). But other companies, on trying to invent their own treatments for high blood pressure, discovered nothing. Science, being unpredictable, does not always yield results.

By definition, research is unpredictable because if it were predictable it would not be research. And because economic growth is ultimately based on innovation – managerial as well as scientific – it too is unpredictable. Managerial innovations often fail, and marketing, too, is unpredictable. The slogan 'You're never alone with a Strand' killed a brand that was thereafter stuck with the image of Freddy No-Friends. As Lord Leverhulme of Unilever used to say, half of his advertising was wasted, but he never knew which half.

Because economic growth is based on an unpredictable entity, namely innovation, Alchian showed that only the evolutionary model explained growth. In a competitive market companies must innovate but, innovation being unpredictable, the inventions that companies bring to the market will be unpredictable. Purchasers will then select among them, and only some products will flourish.

We thus see great similarities between economic growth under markets and evolution by natural selection. In Nature, a system evolves by natural selection if: siblings vary randomly, offspring are abundant, they are selected among, and their descendants inherit their traits. In markets economic growth depends on innovations being created randomly and abundantly by investment in R&D departments and other innovators, on their being marketed, and on selection being made by customers. One unexpected similarity between genetics and companies is inheritance. Companies inherit their differences.

Alfred Chandler's *Scale and Scope: The Dynamics of Industrial Capitalism* reported that firms differ in their cultures. They are created in the images of their founders (some firms are kind, others disciplinary; some firms foster research, others distrust it, etc., etc.), and firms imprint on their cultures, retaining them as a corporate inheritance. Research-friendly or kind firms remain research-friendly and kind even unto the seventh generation of managers. But in *An Evolutionary Theory of Economic Change* (1982), the economists Richard Nelson and Sidney Winter introduced a twist: firms are not limited to Darwinian selection. They embrace Lamarckian or learned evolution as well. Even though firms will generally retain their core cultures, they will learn techniques from each other, and they will learn from the market. Indeed, markets can be too Lamarckian, with players copying each other blindly like herd animals through nonsenses like the dot.com boom and bust.

* * *

To conclude, therefore, evolution by natural selection and economic growth by markets are similar and niche-based. And, Lamarckian or

Darwinian, the survival of the fittest is the invisible hand: as Keynes said, *The Origin of Species* is 'economics couched in scientific language'[10]. The economies that, in practice, approximate most closely in *one* respect (large number of firms) to so-called perfect markets are the ones that grow fastest, but that does not endorse perfect markets: economic growth in the real world is maximized by a large number of firms to maximize competition to maximize R&D to maximise product differentiation and niche creation and the setting of monopolistic prices; while theoretical perfect markets require a large number of firms to minimize product differentiation in favour of product interchangeability and the taking of common prices. So, though perfect markets in theory and economic growth in practice are both optimized by a large number of firms, that certainly does not prove that the theory is correct! Perfect markets, in short, are a fiction, as are the stories of 'market failure' in science that they breed.

Actually, the world of economics has yet to address science properly. Some biologists joke that a rat is an animal that, on being injected with a drug, produces a scientific paper, and a mathematical model is the economists' equivalent. Consequently Romer's formulae have opened a floodgate of papers, and some economists have even claimed that too much science is funded by the market. Since a patented monopoly is so valuable, will companies not duplicate research to 'business-steal' or 'patent-race'? And does not new technology displace old technology before it has depreciated? Should not, therefore, the government fund science to reduce the amount of industrial research?

No one knows the answer to these theoretical questions, because no economist has mathematically modelled the invisible colleges. In practice, though, there is no systematic evidence showing that the government funding of science stimulates economic or productivity growth, which says something important.

16

Let's Abolish Patents

Q: Who were the first people to crash an aeroplane?
A: The Wrong Brothers. (Old joke.)

The twentieth century opened with three memorable technological advances. In 1901 Guglielmo Marconi sent the first transatlantic radio signal, in 1903 Henry Ford launched his production line to bring cars to the masses, and in the same year Orville and Wilbur Wright flew the first manned heavier-than-air powered aircraft, *Flyer 1*. But what is now forgotten is how all three advances were handicapped by patent fights. Marconi fought Nikola Tesla through the courts for no fewer than twenty-nine years (before losing), while Henry Ford fought the patent on the motor car held by the Association of Licensed Automobile Manufacturers (a cartel of bespoke manufacturers that refused to licence a mass producer like Ford) through another morass of courts, winning only in 1911. Meanwhile, the Wright brothers patented the aeroplane, which was the biggest mistake of their lives.

The Wright brothers were bicycle manufacturers from Dayton, Ohio, who invented the aeroplane in their spare time. They were amateurs. The person who felt he should have invented the aeroplane was a grander figure, Samuel Pierpoint Langley, the director of the Smithsonian Institution. Since 1885 he had been trying to fly his own planes, *Aerodromes 1* to *6*, yet each had crashed on take-off into the Potomac River, over which Langley launched his *Aerodromes* to allow his pilots a chance of survival. A reporter described the crash of 7 October 1903 as *Aerodrome 6* 'entering the Potomac like a handful of mortar'. Yet on 17 December Orville and Wilbur Wright took off near Kitty Hawk, North Carolina.

The Wright brothers had financed their own R&D (a mere $1,000) whereas the federal government had provided Langley with a grant of no less than $73,000. The federal government then, of course, funded only military and agricultural research, but Langley had exploited the Spanish–American War of 1898 to persuade Congress to finance him, only to have created, in the disenchanted words of one Representative, 'a mud duck'. Both the government and the Smithsonian were therefore chagrined by the Wrights' success.

But the chagrin was soon aggravated by the Wrights' patents because, after the brothers' success with *Flyer 1*, other American aviators, including Glenn Curtiss, soon built their own planes. But each time Curtiss or any other aviator took to the American skies, the Wrights sued for patent infringement. Official America took Curtiss's side, and in court the Smithsonian and the relevant federal government agencies claimed, falsely, that the Smithsonian's *Aerodrome* had flown first. The Smithsonian even got Curtiss to adapt Langley's surviving *Aerodrome* to show it could have flown, and for years the Wrights were reduced to protesting that it was only on being adapted in the light of later experience that *Aerodrome* (nearly) flew. But official America so denigrated the Wright brothers that in 1928, when Orville Wright (the surviving brother) sought a museum for *Flyer 1*, he found no US institution prepared to take it nor one to which he was prepared to donate it. He sent it instead to the British Museum in London. Only after Orville died in 1948 did the Smithsonian ask London for America's plane back – the Smithsonian did not want to give a Wright brother the satisfaction of knowing that it acknowledged his priority.

Yet this unpleasant story was not just one of frustrated *amour propre*: the federal government had legitimate concerns. The aeroplane was of strategic value, and the Europeans (who readily paid the

Wrights' licence fees) were pulling ahead in aeronautics, but the US was threatened with obsolescence because the federal authorities would not pay those fees. It is hard to understand why America could not reach an agreement on licence fees, but the federal government had funded Langley's research, so it did not want to recognize its waste of money, and the Smithsonian colluded with the charade because it needed to sustain the credibility of future government grants.

The impasse, moreover, seems to have had psychological similarities with the Gates/Microsoft antitrust case of the 1990s, in that Washington, D.C. does not like pig-headed, over-mighty subjects who fail to defer to politicians. In contrast, as Craig Venter reported in his 2007 autobiographical *A Life Decoded* (p.139) a smooth entrepreneur like Rick Bourke, when planning a new company, collected together: '. . . three senators: George Mitchell, who was senate leader, and the heads of the budget committee, James R. Sasser from Tennessee and Paul Sarbanes from Maryland.'

If the federal government was determined not to pay the licence fees, it would have been more honest of it to have modified the relevant patent laws. Indeed, in a key episode, the federal government *did*, as a war measure in 1917, revoke the Wrights' patent rights, a revocation that it sustained until 1975. Between 1917 and 1975, therefore, the federal government forced all US aeroplane manufacturers to pool their patents collectively – and the consequence was the vast growth of the US aeroplane industry. Thus we see that the Wright brothers' patents destroyed aeronautical innovation in the US, and that only on their revocation in 1917 did America's planes take off.

PATENTS DO NOT DISCLOSE NEW INFORMATION

The arguments for patents are twofold: that they incentivize research, and they promote disclosure. Both claims are false. As we saw in the previous chapter, companies have to innovate or they die, so they are already fully incentivized. The company that eschews R&D goes bust. As we also saw, the argument that companies will underinvest in research because they cannot capture all the benefits of their research is also false, because it is based on the misleading 'perfect' market model. In the real world, companies' only legitimate aspiration is to optimize – not monopolize – the profits from their discoveries.

Industrial secrecy, moreover, is a myth: company scientists belong to an invisible college, and they trade information, which they therefore release speedily. In a survey of 100 US firms across a range of manufacturing industries, Edwin Mansfield and his colleagues found that:

> Information concerning development decisions is generally in the hands of rivals within about 12 to 18 months on average, and information containing the detailed nature and operation of a new product or process generally leaks out within a year.
>
> 'Imitation Costs and Patents: An Empirical Study',
> in *Economic Journal*, 1981, 91:907–18

There is, therefore, no market failure in industrial science: a delay of a year to eighteen months is enough to allow an innovator to consolidate their first-mover commercial lead, while being short enough to benefit society by empowering second-movers.

In any case, relatively few innovations are actually patented. In a systematic review of industry, Mansfield and his colleagues also found that: 'Patent protection did not seem essential for the development and introduction of at least three fourths of innovations.' They further found that patents increased the costs of imitation only to a minor degree. Empirical studies show repeatedly that *the* great defence of technological monopoly is 'first-moving' – consistently providing the novel product that everyone wants to buy and for which consumers will pay premium prices, which therefore rewards the first-mover with the profits by which to invest in the next innovation:

> The picture is striking. For new processes, patents were generally rated [by 650 executives across a range of US industries] the least effective of the mechanisms of appropriation . . . Lead time, learning curves, and sales and service efforts were regarded as substantially more effective than patents in protecting products.
>
> R.C. Levin, A.K. Klevorick, R.R. Nelson and S.G. Winter,
> 'Appropriating the Returns from Industrial R&D' in *Brookings
> Papers on Economic Activity*, 1987, 3:783–820

The free market in ideas is, therefore, benign, because nothing more benefits society than a stream of first-mover innovations.

POOLING RESEARCH

Sometimes, when anti-trust law allows, the invisible colleges can become more visible, and companies can openly generate public data jointly. In 1998, for example, most of the world's major drug companies joined the Wellcome Trust, the Cold Spring Harbor Lab and the NIH jointly to fund with $75 million the generation of an SNP map. This map extracts valuable data from the human genome project, and all 150,000 SNPs will be published freely. Patents on them will be filed, but not issued, to keep them in the public domain.

Equally, the Gigascale Center, based on the Berkeley campus in

Silicon Valley, aspires to place a billion transistors on a single chip. The Center involves the Department of Defense, twenty-two universities and some two dozen semiconductor firms, including Intel, Motorola and Advanced Micro Devices. The Gigascale Center's annual budget is $8.3 million, it employs over a hundred researchers, and it publishes freely. In the words of David Kirp, a professor of public policy at Berkeley:

> By subsidizing the Center's research, the sponsoring semiconductor companies have in essence paid membership dues to join an intellectual club where there are no secrets – a scientific community where everyone has a chance to learn from everyone else.
>
> D.L. Kirp, *Shakespeare, Einstein and the Bottom Line*,
> Harvard, 2003, p.209

An intellectual club where there are no secrets and whose sponsoring companies have paid membership dues is, essentially, an invisible college; so we see that the SNP consortium and the Gigascale Center are only public expressions of the reality that company scientists cooperate. Those particular consortia could afford to be open because, by collaborating with charities and government agencies, they bypassed anti-trust law – but had charities and government agencies not been involved, those companies' scientists would have found other, discreet, ways of cooperating.

And a good thing too, because anti-trust law is, in the context of research, an ass: to use economists' language, companies that cooperate in their research will attain a Pareto optimum of maximum benefit to society because each company, by sharing its scientific knowledge, will ensure that no secrets are kept, thus optimizing the perfection of knowledge; but also each company, having identified the technology of its choice, will compete in the market with its different commercial technologies, thus optimizing competition in the market-place. And this optimal order of maximal competition and maximal knowledge-sharing will have been achieved spontaneously, without patents or government direction.

PATENTS DAMAGE YOUR WEALTH AND YOUR HEALTH

People take out patents to inhibit the competition, so perhaps it is right that so many patent-holders suffer. As Karl Marx said, capitalists seek to close markets, not open them, and the Wright brothers were more interested in their profits than in fostering American aviation. In the absence of patent laws, the two sets of interest would have coincided, but patent law set them in opposition.

In any case, the Wright brothers did not invent the aeroplane in a vacuum: in a letter of 1899 the brothers had written to Langley asking for information. And the irony is that the Wrights would have produced *Flyer 1* in the absence of patent laws. They did not create powered flight for profit; instead, like most great researchers, they were driven by the love of discovery. The patent laws being in existence, however, they naturally exploited them, but their persistent litigation brought them only unhappiness.

Indeed, from about 1908 the Wright brothers produced no more innovations, they simply fought patent case after patent case. They won their cases at vast expense, but, ironically, they would have made real money if, instead, they had focused on growing their very considerable first-mover advantage. (But they were stubborn brothers. They never married and when in 1925 their sister, who was their housekeeper, fell in love and left home to marry, Orville, the surviving brother, never forgave her 'treachery', and he never spoke to her again until she was literally on her deathbed.)

Nor were the Wrights alone in their patent defence-induced misery: Eli Whitney (cotton gin), John Kay (flying shuttle), Jonathan Hornblower (double-chambered steam engine), Charles Goodyear (rubber vulcanization) and the Fourdrinier brothers (mechanical papermaking) were but some of the many other inventors who ruined themselves in defending their patents. Whitney, as an industrial researcher, learned his lesson, and he launched a second, successful, innovative career as a manufacturer of firearms – without writing another patent. How wise Jonas Salk was when, understanding that academic scientists should seek esteem via reputation, not money, he dismissed the idea of patenting the polio vaccine as 'like patenting the sun'. Many scientists continue to make their work free, one example being Linux, the open-source software provided by Linus Torvalds and his merry band of fellow idealists. Open source, of course, often outperforms the proprietary sector because of the contributions of the virtual community. Open source also challenges the assumption that innovators require IPR to be incentivized.

Too often, though, the patent holders win, to society's disadvantage. Consider Watt's steam engine. Watt conceived the idea in 1764, and he and his partner Matthew Boulton obtained a patent by private Act of Parliament that extended from 1769 to 1800. Edmund Burke MP protested in Parliament that Watt's patent would damage the British economy, and he was right. In their book *Against Intellectual Monopoly* (which is freely available on the web, of course) two American economists, Michele Boldrin (Washington University at St Louis) and David Levine (UCLA), show how Watt's patent damaged the development of the steam engine. So between 1769 and 1800 (the thirty-one years of Watt's patent) total steam horse power in the UK rose from 5,000 to only 35,000, but after 1800 it accelerated, and by 1815 it had risen to 100,000 and by 1830 to 160,000.

Between 1769 and 1800 the numbers of steam engines in the UK rose from 510 to 2,250 but, remarkably, almost all the newly installed engines were old-fashioned Newcomens that were not covered by Watt's patents. Only 449 were Watt engines, which industrialists avoided because the licence fees were so onerous. Moreover, the fuel efficiency of steam engines did not rise between 1769 and 1800, but it rose five-fold between 1810 and 1835. Why? Because only then did people start to use the high-pressure steam engines that Trevithick had invented but that Watt had blocked by his patents. Watt's patents, therefore, did little but delay by two or three decades the replacement of inefficient Newcomen engines by efficient Trevithick engines.

We thus see the moral argument against intellectual property. A piece of tangible property such as a house can be occupied by only one person, so we need property rights to stop people fighting over it. But an idea can be used by an infinity of people, so by inventing intellectual property we create a potential for obstruction and legal conflict. But entrepreneurs have sought monopoly for millennia (remember Herbert the Dean and his windmill in Chapter 9), and they will always find clever arguments to justify it. As Adam Smith said in the Wealth of Nations: 'Men of the same trade seldom meet together . . . but the conversation turns into a conspiracy against the public'. It is the politicians' job to block the conspiracy that is patents.

And thus we also see that Douglass North, in his claim that 'the failure to develop systematic property rights in innovation up until fairly modern times was a major source of the slow pace of technological change', has confused cause and effect. It was the growth of property rights in tangibles, and the growth of the appropriate commercial and scientific institutions and cultures, that fostered innovation. Successful innovators have subsequently lobbied for IPR, but fat cats will always try to monopolize the cream: 'capitalists seek not to open markets but to close them.'

PATENTS: THE EMPIRICAL EVIDENCE

The argument against patents can be bolstered by some emprical facts. Eric Schiff, the economic historian, reported in his 1971 book *Industrialization without National Patents* that neither Switzerland (from 1850) nor the Netherlands (from 1868) enforced patent laws. But that free-for-all worried Germany which, being both countries' major trading partner, feared they might steal its research. So it pressured them into introducing and enforcing patent laws, which Switzerland did in 1907 and the Netherlands did in 1912.

What were the consequences of these changes in patent law? As Eric Schiff reported, their introduction in 1907 made no difference to the underlying rate of Swiss economic growth, nor did it affect the Dutch after 1912. Nor did the lack of patent protection inhibit those countries' technical fertility. To take just one fourteen-year period in Switzerland, in 1866 Henri Nestlé developed a formula milk for infants, in 1869 Julius Maggi invented powdered soup, in 1875 Daniel Peter invented milk chocolate, and in 1879 Rudolf Lindt developed chocolat fondant. Those names still live in the companies that flourished patent-free on the back of those technologies.

The absence of patents did, of course, facilitate the import of technology. The Dutch precursor of Unilever created margarine during the 1870s by exploiting, licence-free, a French patent, and during the 1890s Gerard Philips in the Netherlands built his company by manufacturing light bulbs without paying Edison a licence fee (Philips's first big order being, ironically, to light a candle-manufacturing factory). But those companies have since produced important innovations of their own, thus showing how their intensification of research has benefited society at large.

One paradoxical example of the value of that research intensification was provided by the Swiss drug company CIBA (now part of Novartis), which was founded in 1869 in Basle to exploit, licence-free, the discovery of mauve and other aniline dyes that William Perkin had made in London two years earlier. Yet, as Simon Garfield showed in his 2000 book *Mauve*, it was the patent-free exploitation of mauve by continental chemists that saved Perkin's own business! Perkin could not find investors in Britain, who did not believe that mauve was a profitable colour, and only after the Europeans started copying Perkin's technology to produce mauve gowns for Parisians (who loved them) did the ladies of London demand their own, thus impelling British investors into funding Perkin's business. Thus we see, yet again, how competition – one that patents are designed to inhibit – stimulates R&D.

But patents are a weapon the rich can aim at the poor because, inevitably, the rich do more science than the poor. The World Trade Organization (WTO) is dominated by the West, and its 1986–93 Uruguay Round established overstrong intellectual property rights for advanced Western countries. As the Economics Nobel laureate Joseph Stiglitz concluded in his 2002 book *Globalization*, the strengthening of IPR during the Uruguay Round helped ensure that 'the result was that some of the poorest countries in the world were actually made worse off'. So, for example, the Uruguay Round prohibited the Third World from manufacturing cheap generic drugs, and the subsequent rise in deaths – 'Millions of people are dying and will die because trade is privileged over human beings,' said James Orbinski of Médecins sans Frontières – proved such a scandal that, in its Doha Development Agenda of 2001, the WTO made concessions, allowing some Third World countries to manufacture cheap generic drugs under compulsory licences.

Alarmed, the corporations have been fighting back. In 1998 Helmut Maucher, CEO of Nestlé, assured the WTO that: 'the protection of intellectual property is essential for economic growth'. Meanwhile, Novartis battled the South African government for three years over the cheap local production of its HIV therapies. And Unilever helps lead Europabio, the powerful lobbyist for even stronger global corporate patents.

These companies are only defending their selfish interests, and discreetly they know what they are doing: their loyalties are reserved for their shareholders, not to society. Less defensible are the useful idiots who defend IPR intellectually. Too many neo-conservatives, in their obsession with property rights, refuse to acknowledge the difference between rights in tangible objects and those in knowledge. So, for example, when Paul Wolfowitz was at the Defense Department, his website boasted that when he had been ambassador to Indonesia between 1986 and 1989 he had been 'a tough negotiator on behalf of American intellectual property owners'. But even Wolfowitz knows what impact such American intellectual property ownership had on Indonesian economic development, and his World Bank site said simply that during his time in Indonesia: 'he was known for reaching out to all elements of society and for his advocacy of reform and political openness.'

Within the developing world, patent protection seems not to promote endogenous private research. Consider agriculture. As we saw in Chapter 15, the economists Carl Pray and Keith Fuglie of the US Department of Agriculture found that private agricultural R&D will flourish within the developing world when the market is free. Changes in IPR, however, make little systematic impact:

> There were substantial changes in IPR policy during this period [of the study] but they were not consistently associated related to changes in research intensity. For example, Malaysia and Thailand made improvements to their patent laws but had declining research intensity.

India and Pakistan, which had very limited IPR changes, had the most
rapid growth in research intensity.

US Department of Agriculture, *Agricultural Economic
Report* No. 805, 2001

Pray and Fuglie noted that, where IPR is weak, companies find
other ways of protecting their innovations, of which the most
important were technology and secrecy. These do not always benefit
society:

Seed companies protect new plant varieties by producing only [sterile]
hybrids. Chemical companies protect new chemicals by keeping their
syntheses secret and by making them hard to copy.

But in practice new seeds (many non-sterile) are always being
developed, and we know how ineffective industrial secrecy is in
reality. Indeed, the two most famous of industrial secrets, the formulae
for Coca-Cola and Kentucky Fried Chicken, have long been known to
their competitors – who keep the secret because it is actually in their
own interests to do so. 'Men of the same trade . . .'

Further, we should note that the abolition of America's aviation
patent pool in 1975 was not forced by any evidence that it was
damaging the US's planes (it wasn't) but by an ideological Justice
Department (it was the Nixon/Ford era) responding to theoretical
concerns over IPR (it was, of course, the wrong economic theory).
Equally, of course, the American car industry took off only because
Ford broke the ALAM (Association for Licensed Automobiles
Manufacturers) patent. And remember that yet another major US
industry, films, took off only after the auteurs had decamped to
California; not, as is commonly assumed, in pursuit of sunshine
('outside' scenes were then often filmed in studios) but to escape

Edison's patents. Edison, who litigated ferociously in defence of his patents on raw film, found it difficult from New Jersey to police film-making in southern California.

DO PATENTS HAVE ANY USES?

Patents do allow science to be traded in the form of licences. In 2000 the National Science Foundation reported that between 1980 and 1998 no fewer than 9,000 major US, European and Japanese firms entered into strategic technology alliances, sometimes leading to unusual newspaper headlines, such as that of 13 August 1997 in *The Times* of London, which described the deal between Apple and Microsoft as the 'Rivalry that ended in friendship'. Meanwhile, IBM invests so much in R&D that for the last twelve years it has been awarded more US patents (3,000 annually) than any other institution. And from its total of 40,000 patents IBM earns over $1 billion annually in license fees, thus accounting for over 2 per cent of all American licence fees.[1]

But are such licence deals really optimal? Japanese patents are weaker than US or European patents, and they provide inventors with less protection, with the result that, under the protection of con-fidentiality, Japanese inventors enter more readily into technology-sharing agreements with companies, thus speeding the dissemination of inventions within Japan and enhancing its productivity, which has in turn helped boost Japan's vast private investment in research.[2]

Although it is supposed that lone inventors need patents to protect their ideas from being stolen by Big Industry, in practice lone inventors can generally approach companies safely:

Very few of those [big firm managers] could recall inventions sub-mitted from individuals or very small firms that had been accepted,

375

although one or two isolated cases were mentioned, but all said they were prepared to welcome promising cases. Most of the inventions submitted are relatively simple-minded, although some show genuine technical expertise or ingenuity, and the main reason for not taking them up are that the idea is old or that it is simply not a commercial proposition.

Christopher Taylor and Aubrey Silberston, *The Economic Impact of the Patent System*, Cambridge University Press, 1973, p.322,

In reality, therefore, few lone inventors have much to offer big companies, which, possessing teams of scientists and having routinized cutting-edge research, can generally beat the tiddlers. And, when lone inventors do pull ahead, they find protection in legally binding confidentiality agreements, which are easy to write and which have become standard practice, and through the fear of:

The adverse publicity that tends to attach to a large company involved in a court action, especially where the opponent is an individual or a very small firm.

C. Taylor and A. Silberston, op. cit., 1973, p.102

Such litigation is, of course, expensive but so too are patents, whose cost can be prohibitive to lone inventors. Frank Whittle, for example, the lone inventor of the jet engine, could not afford to maintain his patents, and they lapsed (though their irrelevance was highlighted by the successful creation of the patentless Power Jet Company in 1936). Indeed, patents are so expensive as to preferentially empower rich companies over lone inventors. A 1996 OECD report showed that 60 per cent of all US patents are filed by fewer than 700 firms.

Venture capital

Contrary to myth, big companies – whose executives have their own vested interests, can often resist (rather than copy) new ideas: famously, James Dyson of the vacuum cleaner could not interest the established companies in his revolutionary design. But lone inventors are empowered by venture capital, which embraces innovation.

The first specialized venture capital company, American Research and Development, was created in 1946 to exploit the ideas that started to spill out of the universities following the inauguration in 1940 of the federal government's funding of academic science, and the venture capital industry has flourished ever since. And because only the inventors know the full, tacit facts of their inventions, they hold the whip hand over venture capitalists. Thus we see how the power of lone inventors *vis à vis* the venture capitalist balances their vulnerability vis à vis the big company, and a socially useful equilibrium is reached that requires no government subsidy.

WHO CARES ABOUT LONE INVENTORS?

Is lone invention so precious that we should care much about it? Since many lone inventors are academics, their defenders invoke a study of Edwin Mansfield's. Mansfield was the scholar who showed that some 10 per cent of industrial innovations arise from recent academic research. Even though he found those innovations to be economically marginal, accounting for only 3 per cent of sales and 1 per cent of the industry's savings, he then performed a famous calculation. He determined the investment in all academic science worldwide and calculated industry's annual return on that investment as being no less than 28 per cent! Therefore, like the good lobbyist that most American academics are, he argued that: 'for policy-makers who must decide how much to invest next year in academic research, this incremental rate of return is of primary significance.'[3]

But Mansfield was simplistic. He assumed that academic science was turned into commercial technology cost free. In fact, academic research and lone invention account for only a small proportion of the total costs of any new product. The development costs will be at least ten times greater, yet even those will be dwarfed a further ten- or hundred-fold by the new manufacturing and distribution costs, marketing and so on. Thus, once the costs of development and production are included, the annual return on academic science or on lone invention is indeed small.

Something else is small: the value of ideas at an early stage. A survey in 2003 by Thomas Astebro of the University of Toronto of 1,091 Canadian inventions revealed that only seventy-five reached the market, of which forty-five lost money.[4] Inventors, therefore, are engaged in a risky tournament where only a tiny minority of players are rewarded handsomely. Thus we see that new ideas are rarely the limiting factor in commercialization. Scientists and inventors, raised on Bacon's linear model, assume that the rate of discovery determines the rate of wealth creation, but actually development, production, retraining, retooling, marketing and investment are the limiters in commerce. Consequently, in the commercialization of ideas, a recurring theme recurs – community. Only the inventor who is integrated within a network of commercial colleagues will realize their invention. The myth is that invention is rare, but actually it can be routinized: the reality is that the well-managed enterprise capable of realizing profits out of new technology is rarer.

The commercialization of research is an unpredictable business. Some inventors (Bill Gates) do inappropriately well, others (the Wright brothers) don't. Such uncertainties upset economists like Richard Nelson, who ask why, in the absence of theorems of welfare economics, the capitalist innovation system should work so well. The answer lies in probability theory. As long as enough trades are being performed and as long as the average balance of advantage is equitable, the mass of trades will yield the social optimum they currently do.

* * *

In conclusion, therefore, patents should be abolished. Not only do the vast majority have no commercial value (over 90 per cent are merely vanity publishing), but the valuable ones inhibit competitors. Only when markets cannot, inherently, be brisk, should patents be retained. One industry only, the pharmaceutical industry, seems to match that criterion. There, government safety regulations are, rightly, so exhaustive, that after its discovery each new product requires up to fifteen years and $900 million in development costs before it can be prescribed clinically. Because pharmaceutical imitation must be so much cheaper than innovation, pharmaceutical innovators do require the legal protection of patents.

But only in the pharmaceutical industry is the empirical evidence persuasive: the stronger the patent laws, the more the pharmaceutical companies invest in R&D.[5] The lack of such evidence in other industries is telling.

17

In Which Adam Smith is Vindicated by the OECD

'What is to be done?' Vladimir Ilyich Lenin, 1902

This book opened with Sir Francis Bacon's suggestion that science is a public good and that, in consequence, it needs public funding. In the spirit of Adam Smith I countered with the suggestion that science is actually an invisible college good and that, in consequence, it needs no public funding. These are fine assertions, and they can be tested by 'crowding out' and 'crowding in'. If science is an invisible college good, then it will behave as a private good and its subsidy by government will 'crowd out' its private supply. But if science is a public good, then its subsidy by government will increase the total amount available, and it might even 'crowd in' its private provision. What do these terms mean?

Consider a typical private good, the motor car. We each buy our own. But what would happen if the government regularly supplied us with free cars? Most of us would buy fewer, and the private retail car dealerships would flounder. The government will have crowded out the private market.

Now consider a public good such as a road. The government supplies roads for free (at least at the point of usage). Consequently, people buy private cars to use the public roads. The provision of a public good like a road, therefore, 'crowds in' the car industry, because people will buy private goods to use a free public one.

By analogy, therefore, if science were a private good like a car, its provision by government should 'crowd out' its private funding, because companies would save money and use the government's free science. But if science were a public good like a road, its provision by government should 'crowd in' its private funding because companies would invest in the research opportunities offered by the new free knowledge.

Because different governments follow different policies at different times, the economists can determine if subsidies for science have crowded in or out its private supply. They can therefore tell us if it is a public good or not. (Actually, analysis can be complex. Consider, for example, the government provision of free family cars, which might crowd out people's buying of them. However, some people might then buy sports cars as luxuries they could otherwise not have afforded, so the government provision of free family cars might boost the manufacture of boys' toys. Is that crowding out or in? We must be careful how we interpret these surveys.)

Over thirty-five separate surveys have now been published on whether government money for research crowds private money in or out. When governments give money to companies for science (incredibly, that's what governments do) or when governments supply public science in the shape of universities and research councils, does the private sector then fund less or more science?

CROWDING IN AND OUT: INDIVIDUAL FIRMS

Most surveys have focused on the responses of individual firms, and their conclusions are finely balanced. In the words of a comprehensive review:

There seems to exist near equality between the number of studies that obtain a substitution [crowding out] and an additionality [crowding in] result . . . it is obvious that differences in the institutional environment play a decisive role.

Wim Meeusen and Wim Janssens, 'Substitution versus Additionality', *CESIT Discussion Paper 2001/01*, 2001

Surveys show, therefore, that government money for company research sometimes crowds in, and sometimes crowds out, private money. Although that contradictory result seems unhelpful, it actually shows that science must be a private good (such as an invisible college good) for two reasons.

First, the logic flows only one way. Public goods do not crowd out, but private goods can crowd in or out, depending, in the words of Meeusen and Janssens, on the 'institutional environment' (such that if governments provide free family cars, they will crowd out private family cars but crowd in private sports cars). I will not here describe the many different ways by which governments have subsidized company science, ranging from grants to companies to the fostering of consortia to the provision of facilities, but the fact that company science has responded to public subsidy variably, much as cars would have done, suggests that science behaves as a private or invisible college good.

But regardless of the logic, the studies that did apparently show crowding in were generally flawed, because of the inappropriate assumptions of the economists. This has been a surprising story. Let me tell it as the economists tell it, as a tale of two companies, one of which (Company A) increases its research budgets, while another (Company B) does not.

Company A decides to increase its research budgets. What will it do? Obviously, it will apply for private money in the form of bank loans and venture capital, and, because government grants are available, it will apply for them also. Company A, therefore, will both increase its private research budgets and also take government research money. But Company B decides that it need not expand its research. It wants to increase its marketing budget instead. So it does not apply for government research grants nor for bank loans or venture capital money for research.

The facts are obvious: companies apply for government and private money together or not at all. Clearly, therefore, there will be a strong correlation between those companies that obtain both private and government funding for research and those that get neither. The facts indeed show that. But the economists work to two false assumptions. First, they assume that all companies seek to increase their research budgets at all times (a bizarre assumption), and second, they pre-assume that science is a public good – but that is the hypothesis they are testing!

So they explain Company A (the company that increases public and private funding for R&D together) as proving that, having got public funding, the public funding must have crowded in the private funding; but that is false. Company A had already decided to increase its research budgets, and had government money not been available it would simply have obtained it from the usual private sources alone.

And the economists explain Company B (the company that got no R&D funding increases) as having failed to obtain private money because it failed to get a government grant. But that is false. Company B had already decided not to increase its research budgets, and it got no government money because it did not apply for any.

Because of these false assumptions, (very) circular arguments and other methodological imponderables, even Professor Paul David of Stanford University, a prominent economist of science and passionate lobbyist for the government funding of science, concluded in his recent joint review of the field that, 'the findings overall are ambivalent and the existing literature as a whole is subject to criticism,'[1] which represents a remarkable admission by one of the leaders of the current orthodoxy that, at the company level, science has not been shown to behave as a public good.

Surveys show, incidentally, that the only way a government can be sure of increasing a firm's expenditure on R&D is by providing it with tax credits. Yet even that route is flawed, because the firms whose R&D is most vulnerable – the small start-ups with no revenue – pay, of course, little tax, so they barely benefit. Moreover, when tax credits are available, the accountants from the big companies pile in all sorts of inflated and mythical expenditures under 'R&D' to claim the benefits, regardless of how much actual R&D they may do.

Even more worryingly, as two leaders in this field, Bronwyn Hall and John Van Reenen of the Institute of Fiscal Studies, have shown, tax breaks really do increase firms' real expenditure on R&D. But do we actually want more industrial research? Here is a quote from their paper:

> Lowering the cost of research might cause the firm to do too much. Even though the tax credit induces more industrial R&D than the lost tax revenue, it would not be a good idea, because one could have spent that tax revenue on some other activity, which had a higher social return.
>
> 'How Effective are Fiscal Incentives for R&D?'
> in *Research Policy*, 2000, 29:449–70

Hall and Van Reenen, therefore, worry that government incentives might cause a firm to do too much research. They're right, of course, because companies certainly can overspend on research. Research, on average, brings in private returns of 20 per cent annually, but companies can often earn greater returns – to themselves and therefore to society at large – by spending on equipment, staff, training or marketing.

Hall and Van Reenen, however, take comfort in the 'available evidence on the social return to R&D', which, they believe, suggests that every dollar spent on research by a private company benefits the rest of society several times more. Unfortunately, though, we now know that there are no social returns to R&D; there are only collegiate returns to other researchers, which have to be earned. (If there were social returns on R&D, those returns would not respect national borders, and Africa would be as rich as America. Africa is poor because it hasn't joined the invisible college, not because knowledge cannot cross the Atlantic.)

Over-investing in R&D, incidentally, has long been a national problem. The governments of countries such as the USSR and post-war India invested hugely in science – to no economic benefit. Those countries were poor, and they needed roads and markets, not particle accelerators.

CROWDING IN AND OUT: INDUSTRIAL SECTORS

Some economists have looked at crowding in and out for whole industries: if government supports steel research, does the whole industry consequently raise its research spend or not? Again, some studies report crowding out and others crowding in. And yet again, as Professor Paul David continues honestly to admit, the studies that do apparently report crowding in at the whole-industry level may be flawed: they may actually only be reporting 'the correlated effects of other macroeconomic variables'. Let me explain what those 'macro-economic variables' are, using the USA to explain the technical terms[2] (the USA being big, its research is largely typical of OECD countries).

R&D

Research comes in different guises, but 'R&D' refers to all 'research and development', wherever it is conducted. R&D budgets for the lead countries are large. The total US R&D budget for 2000 was $264 billion, which was 2.7 per cent of GDP. That share of GDP is typical of the lead countries.

Civil versus Military R&D

Some research is military. In 2000 the US spent $42 billion on defence R&D, which was 0.4 per cent of GDP and 16 per cent of its total R&D budget. Other lead countries, including the UK, France, Sweden and Russia, also spend up to 0.5 per cent of their GDPs on defence research. But other lead countries, including Germany, Austria, Italy and Japan, spend only tiny amounts.

The great determinant of military R&D is history. Those countries that won the Second World War or that, like Sweden, believe their policy of armed neutrality protected them, are comfortable with their military and will spend money on its research. Those that lost aren't and don't. Either way, the economic impact is small – dollar for dollar, defence R&D yields less than 10 per cent of the economic benefit of civil R&D.[3]

Civil R&D

This represents all non-military research. The US's civil R&D budget for 2000 was $222 billion, and 80 per cent of that was 'D' or 'development', namely applied science taking place in industry. It is as often performed by engineers as by actual scientists. Most industrial applied science in the US today is funded by industry itself. Individual company R&D budgets can be huge – Microsoft's is $4.4 billion, and GlaxoSmithKline's is $2.4 billion – but those are only the most prominent of a large number of companies whose R&D budgets exceed $1 billion.

Research

The remaining 20 per cent of the US's civil R&D budget is pure 'R', namely basic science performed by actual scientists. Half of that 20 per cent is funded by the federal government ($22 billion, mainly within the universities) and the other half is funded privately, either by industry ($15 billion) or by foundations ($7 billion). The federal government, therefore, pays for only half of all basic science in the US (the state governments' contribution being negligible).

Some 'macroeconomic variables'

Why does America spend $222 billion on civil R&D? Why not $22 or $2,222 billion? What, indeed, determines how much research, and of what quality, any economy produces? Is there a 'right' amount of research for an economy, or do countries simply spend whatever the politicians and senior industrialists decide one morning?

The best answer is determined empirically: national expenditure on research is determined by national income. The greater a nation's GDP per capita, the greater is its output of basic science, the greater is the quality of its science, and the greater is the share of GDP it dedicates to civil R&D. Figure 17.1 illustrates the point: the quality of a nation's science can be judged by the number of times its research papers are cited by other research papers, and Figure 17.1 shows that the quality of a nation's science correlates closely with its national income per capita. As does quantity: a graph correlating the number of scientific papers published by lead countries per capita of population against GDP per capita looks similar.[4]

Although national statistics on science are collected on the basis of output (numbers and citations of papers), national statistics on R&D are collected on the basis of input (as its share of national GDP), and it can be shown that the percentage of GDP an OECD nation spends on civil R&D also correlates well with its income: thus a rich country like

Switzerland spends up to 3 per cent of its GDP on civil R&D, while a poor one like Portugal spends only some 0.3 per cent.

Figure 17.1 GDP per capita and relative citation rates

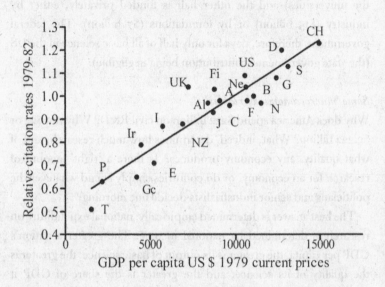

Such graphs establish the remarkable efficiency by which the market economies titrate the needs of innovation and of copying: only rich countries do research; poor economies enrich themselves largely by copying the techniques of rich ones. The poorer an economy, the more backward is its technology, so the simpler is the technology it needs to copy (reliable roads would alone make a big difference to many poor countries). Consequently, a poor economy needs do little cutting-edge research to copy. But rich economies are, by definition, technologically advanced, so they can enrich themselves further only by innovation or by copying sophisticated technology. To join the relevant sophisticated invisible colleges, therefore, their own research needs to be both ample and sophisticated. Rich economies thus need to spend generously on research and, in one of the market's virtuous circles, they can afford to.

One unexpected consequence flows from these facts: because, under convergence, poor countries can grow faster than rich ones, we see the fastest economic growth rates in the countries that do least research! But since the poor countries are largely only copying, that is not so surprising.

SCIENCE SHOWS DECREASING RETURNS IN ITS INCREASING RETURNS

Another unexpected conclusion that flows from these findings is that research budgets grow at double the rate of GDP. When a country doubles its national income, its share of GDP spent on R&D also doubles, which means that its spending on R&D goes up four-fold. Thus lead economies double their incomes every twenty-five or so years, but their research budgets double every twelve to fifteen years. Whether measured by the numbers of scientists, the numbers of journals or papers, or the size of their budgets, the doubling time of science is around twelve to fifteen years. By 1750, for example, only ten scientific journals had been founded worldwide, but by 1850 that figure stood at one thousand, and by 1950 at one hundred thousand.[5]

Indeed, so swift is the doubling time of science that Derek de Solla Price, the physicist who first noted it, remarked that the statement 'over 85 per cent of all the scientists who have ever lived are alive today' has been true for every year since 1750, and it remains true today. (Price, a great wordsmith, also coined the term 'brain drain'.)

Because the share of GDP devoted to research rises with GDP, we must assume decreasing returns in science's increasing returns. What does this mean? Well, we know that new science creates wealth, but each new piece of science apparently creates less wealth on average than the previous one; so, to maintain steady rates of economic growth, more and more science needs to be performed. So science does increase returns, but at ever diminishing rates – that is, there are decreasing returns to science's increasing returns.

This finding has dismayed the economists who, following Paul Romer's claim in his theory of endogenous growth that 'possibilities do not add up, they multiply', had assumed that every new discovery would yield novel combinations of knowledge that would yield ever-increasing rates of economic growth. As William Easterly wrote in his 2002 book *The Elusive Quest for Growth*: 'The more existing knowledge there is, the higher is the return to each new bit of knowledge.'

But the reality is the opposite, namely that research demonstrates diminishing returns in its increasing returns. Why? One reason may be that research gets ever more difficult: inventing the steam engine was easier than cloning genes. If the easy research has, therefore, already been done and only the difficult stuff is left, each new piece of research is going to cost more to complete. But another reason for diminishing returns in science may be that each new bit of knowledge provides, in combination with the existing ones, such a myriad of possible combinations that it becomes hard to identify the potentially useful ones. Consider a pack of cards. It contains only fifty-two cards, but that provides 52×51×50×49 etc. possible combinations of cards. At 80,000,000,000,000,000,000,000,000,000,000,000,000,000,000,000, 000,000,000,000,000,000,000,000,000 possible combinations, not many identical hands have ever been dealt. Science contains more than fifty-two observations, and if innovation comprises the juxtaposition of novel combinations, just think of the potential combinations the scientist has to winnow! No wonder each new piece of science yields fewer returns.

We can conclude, therefore, that economic growth rates will eventually plateau. If R&D budgets need to double every twelve to fifteen years to allow GDPs to double every twenty-five to thirty years, then one day so much of the population will be researchers that we will be forced to restrict further growth in R&D because of the competition for resources for health, education, welfare, transport and other services.

When will R&D budgets plateau? The lead economies seem already to be stuck at 3 per cent of GDP, which has not grown much for two or so decades. But that apparent plateau is deceptive. In 1964 the US federal government, for example, paid for 67 per cent of all US R&D, but by 2000 that had fallen to 26 per cent.[6] So beneath the plateau for total R&D we've seen a huge rise in business-funded R&D. This crowding out of state-funded by business-funded civil R&D is taking place across all the countries of the OECD. And because (as the OECD report quoted in Chapter 1 found) private R&D is the only R&D of economic value, we are therefore only halfway on the rise to its plateau. I suspect that civil R&D budgets will rise up to 6 per cent of GDP before they plateau. That gives us a few more decades yet of economic growth!

CROWDING IN AND OUT: THE NATIONAL PICTURE

Because research budgets rise inexorably, we cannot ask if government subsidies for a whole industry's research stimulate that industry's private funding. If government subsidizes an industry's research and if that sector spends more private money on research, the reality is that it would probably have increased its private budgets anyway because most industries do increase their funding over time. Even if an industry (such as the buggy whip industry) is in decline, who is to say what its research funding would have been whether or not the government subsidizes it? Only Professor David's 'macro-economic variables' can be invoked, not cause and effect.

We therefore face a problem: no scholar has shown how to answer the question of 'does government funding of science crowd in or out its private funding?' All studies, whether at the firm or industrial level, have been flawed. It was to resolve this problem that, previously, I suggested a solution, because there is one set of research budgets that are predictable – the national ones.[7] We know that the quantity and quality of a nation's research is determined primarily by its income per capita.

We can, therefore, determine if, in aggregate, the government funding of research crowds private money in or out. This is because government policies for civil R&D have varied dramatically from country to country. Some countries, such as Switzerland or Japan, have been extraordinarily laissez-faire in civil R&D, with governments supporting only 20 per cent or even less of it (and then mainly for academic science, not industrial research and development). But in other countries such as Australia and New Zealand, governments have paid for up to 80 per cent or even more of companies' R&D (in those countries, companies wanting to do R&D have simply written grant applications, like academics in universities).

I showed earlier that once correction was made for GDP per capita the government support for civil R&D actually lowered the total amount of funding (as is confirmed by a systematic mathematical analysis across all the countries; see the notes to this chapter).

Also using OECD data, but doing a different type of analysis – 'longitudinal' in the jargon – both the OECD itself (see *The Sources of Economic Growth in OECD Countries* which was quoted in Chapter 1) and Walter Park, an economist at the American University in Washington, D.C., concluded that the government funding of

R&D crowded out its private funding nationally. In Park's words, reviewing the major OECD countries for the years 1970–87: 'Once private research is explicitly controlled for, the direct effect of public research is weakly negative, as might be the case if public research spending has crowding-out effects which adversely affect private output growth.'[8]

The title of this chapter speaks of 'vindication' because, when I first proposed that the OECD data suggested that the government funding for R&D crowded out its private funding, Professor David (my enemy) accused me of 'flawed economic logic' and of 'telling of stories that blatantly disregard the evidence'[9], so the later work of Professor Park and of the OECD was indeed vindicative.

Crowding out, moreover, can be confirmed by the stories of individual countries. Consider Britain. Between 1966 and 1974 the British government increased its annual budget for civil R&D from £2 to £3 billion. During that time, the private sector's expenditure on R&D did not rise. But in 1974, responding to the failure of Harold Wilson's White Heat of the Technological Revolution, Westminster froze its budgets. Over the next two decades (to near-universal predictions of national decline) the government's budgets for civil R&D did not rise, yet the private sector increased its annual budgets from £5 billion to £8 billion, thus achieving a greater rate of increase than had the government earlier.[10] We can see, therefore, that the years 1966–74 had simply been years of crowding out, to the detriment of the economy.

Or consider Singapore. That country had long enjoyed huge increases in its national budgets for civil R&D, which rose from S$70 million in 1978 to S$1 billion in 1992, but those increases had been privately funded. In 1991, however, the government created the National Science and Technology Board to subsidize civil R&D by S$2 billion over the next five years. Whereupon the share of GDP dedicated to civil R&D, which had been on track to reach 2 per cent, promptly plateaued at 1.1 per cent GDP![11] Crowding out indeed.

So few people understand crowding out that when I explained that the disappointment had been caused by the Singapore government's own subsidies, its spokesman replied that because both government and private funding had risen in absolute terms after 1991, even if at a disappointing rate, then the government funding must have stimulated the private funding.[12] This bizarre analysis, which ignores his own country's immediate record, shows how economists are persistently blinded to empirical reality by their theories of public goods. It's like the stories of Company A and B all over again.

Until recently even the OECD failed to understand crowding out. As recently as 1991 the OECD criticized the Swiss government for subsidizing its national civil R&D less generously than any other country in the world – when Switzerland as a nation was spending more per capita than any other on civil R&D.[13] To be fair to the OCED, it now understands crowding out.

Or consider the United States. Following Franklin D. Roosevelt's election in 1932, the federal government's budgets for science nearly halved, and, as we know, the private sector responded by increasing its research budgets. Between 1931 and 1938 the numbers of industrial labs in the United States rose from 1,600 to 2,200, their research personnel rose from 30,000 to 40,000, and their annual expenditure rose from $120 million to $175 million.[14] That rise must have owed much to the newly cheapened cost of labour, but crowding out must have played an important role too.

R&D provides a good investment in bad times as well as good. Between 1997 and 2003 the British FTSE 100 Index fell by 15 per cent, reflecting the bursting of the dot.gone bubble, but as the DTI's 2003 Scorecard showed, the share price for R&D-intensive companies – those that spent more than 4 per cent of sales on R&D – rose by 30 per cent.

The mechanisms of crowding out are not hard to find. The obvious one is that government grants to companies allow them to reduce their own R&D budgets. A more subtle one is that the major motive for the industrial funding of science is competition, but many government schemes – predicated on the false assumption that science, being a public good, is damaged by competition – seek to minimize competition and to maximize cooperation. When, therefore, companies are lured by government money into collaborative research with erstwhile competitors, they will spend less on science. So, for example, the companies that entered the federal government's Sematech consortium of semiconductor research ended up spending less in total on research.[15]

Moreover, investment in science is often led by researchers scenting new opportunities generated by their research, and if scientists have been lured into government labs, they will cease to identify investment opportunities for private industry, whose R&D budgets will therefore fall. Charities, too, will desist from funding science if they feel the government is active. As Martha Peck, the executive director of the US Burroughs Wellcome Fund (endowment $600 million) said in 1993: 'We've seen foundations turn away from research toward support for social and community issues. The perception has been that science is getting from other sources [i.e., the government] the kind of funding it needs.'[16]

Further, if companies have been led by government grants into non-commercial projects, such as supersonic flight, the commercial spin-offs will be few and so new R&D funding opportunities will be equally few.

EVEN IF R&D CROWDS OUT, WILL BASIC SCIENCE?

But will pure science respond to government money in the same way as civil R&D? Will pure science crowd out? There is a lot of evidence to suggest that it will. First, the distinction between pure and applied science was always a myth promoted by the now-discredited linear model. In a survey of biotechnology, for example, Francis Narin, a prominent consultant to the National Science Foundation, found that the average delay between a study appearing in the academic journals and a patent was only four years. In Narin's words: 'The division between biotechnology and science has now almost completely disappeared.'[17]

The division has almost completely disappeared because markets are now too brisk for entrepreneurs to ignore so-called basic science. Consider a recent survey of forty-six major innovations, ranging from antibiotics to zippers, of the last 120 years: in 1895 the average delay between the appearance of an innovation and of a competitive copy was thirty-three years; in 1915 the delay was twenty-four years, in 1935 it was fourteen years, in 1955 it was six years and in 1975 it was only three years. On those trends, an innovator can now expect a competitor to get his or her copy marketed within a year and half.[18] Markets are now so brisk that the entrepreneur who does not invest in pure science is asking for bankruptcy. Where was the Swiss watchmaker when the Japanese exploited liquid crystals to develop the electronic watch? In brisk markets, pure science is applied science, period, because every potential opportunity needs urgent investigation.

The division between pure and applied science was in any case never real but social or economic: applied science was science funded by industry, and pure science was science funded by private means or foundations, but actually each type of science was chiselling away at different faces of the same mountain of ignorance, and only snobbery allowed one to claim superiority over the other. The lobbyists, though, responded predictably to Narin's conclusion that there was no substantial distinction between science and technology: the National Science Foundation and the *New York Times* promptly invoked Narin's conclusion to justify the government funding of academic science when, of course, it establishes the opposite – namely that, with pure and applied science being essentially one and the same, there is no need for the government to fund either.

Moreover, in a fascinating episode from recent British history, we saw the private sector move into university research when the state withdrew partially, thus confirming crowding out in academic science. During the early 1980s Margaret Thatcher's government in Britain reduced its subsidies to universities and their science because UK universities enjoyed better staff:student ratios than in the US or anywhere in the world except for the Netherlands (whose government too was retrenching), because UK science was excellent (look at all its Nobel prizes) but had not delivered economic growth, and because the government had run out of money. Mrs Thatcher's cuts were met with universal cries of woe, but the consequence was that industrial and charitable support for university science doubled! As the Committee of Vice Chancellors and Principals conceded in its 1991 *State of the Universities*: 'Increases in funding from industry and medical charities have more than made up for the lack of government funding over the past decade.'

That sort of crowding out is seen all the time: consider the Search for Extra Terrestrial Intelligence (SETI). In 1975 NASA inaugurated SETI, funding universities and researchers to probe the skies for radio signals from aliens. In 1993, however, on the grounds that it was silly, Congress cancelled it. Did SETI die? Hey, this is America, the land of the free. Outraged by Congress's decision, private SETI groups sprang up all over the USA, and the SETI League now integrates the activities of 1,280 volunteers, each of whom possesses a small, radio-telescope into a coordinated search for extraterrestrial life.

The SETI Institute, moreover, now raises over $7 million a year privately, employing over 120 experts to probe the skies with the Arecibo 300 metre radiotelescope in Puerto Rica. The analysis of radiotelescopic data requires vast computer time and, in a remark-able example of voluntary coordination, the SETI@home project (administrative budget $1 million a year) has recruited over 3.3 million volunteers to link their home computers via the net to the Arecibo telescope. Funding the various SETI projects have been big names, including Paul Allen of Microsoft ($25 million to date) and David Packard of Hewlitt Packard, as well as myriad non-millionaires. Hollywood, too, has provided moral support in the form of the 1997 film *Contact*.

SETI thus confirms that when governments won't fund science – no matter how pure it appears to be – the private sector will. The classic example of that was provided by 'Moonie' Goddard (1882–1945). The myth is that the private sector would never have funded space research – yet it did. Robert Goddard was a professor at Clark College, Massachusetts, who between the wars received $10,000 from the Hodgkins Fund (administered by the Smithsonian) and $100,000 from the Guggenheims to create the first modern rockets. By 1925 he had created the first liquid-fuelled rocket, by 1932 he had developed a gyro stabilizer, and by 1937 one of his rockets had climbed 9,000 feet. So successful were Goddard's rockets that one of NASA's first acts was, in 1960, to pay Goddard's estate $1 million for his 200 or so patents (Wernher von Braun in Germany having earlier copied them for his V1s and V2s without permission).

Goddard died in 1945 (having spent the Second World War improving the bazooka, which he had originally developed during the

First World War) and thereafter the federal government supported von Braun, but Goddard's story illustrates that, had war not intervened, the foundations would have funded the launch of the first artificial satellite. Thereafter, the commercial sector would have funded the further development of radio satellites and the like. Meanwhile, we are seeing the rise of space tourism. Space, the final frontier, would have fallen to the private sector.

Moreover, it could have fallen only to the private sector, because only that sector could have boldly gone there. Space exploration before 1940 was viewed by opinion-formers the way they now view SETI. On the publication of one of Goddard's research papers in 1920, for example, the *New York Times* commented on 13 January that he was an 'absurd' person because: 'Professor Clark with his "chair" at Clark College does not know the relation of action to reaction, and of the need to have something more than a vacuum against which to react . . . he only seems to lack the knowledge ladled out daily in high schools.' When in 1929 one of Goddard's rockets failed, and rose only a few hundred feet, the local newspaper ran the headline 'Moon rocket misses target by 238,799½ miles'. The taxpayer of the day would never have supported Goddard, who openly admitted that he had been inspired by Jules Verne and H.G. Wells. Science fiction indeed!

Much of the best science, on its first appearance, is rejected by the authorities as science fiction. The myth is that the universities and the government's funding bodies are centres of unfettered thought, but in practice – constrained by peer review, peer pressure and other mechanisms of accountability that inhibit iconoclasm – they can too easily propagate the safe instead of the unknown. So, for example, the Medical Research Council in Britain refused to support the *in vitro* fertilization (IVF) research of Patrick Steptoe and Robert Edwards, who had to find private funds (including from their own pockets) to enable it. And, in an echo of Goddard's problems with the *New York*

Times, Steptoe and Edwards were forced, successfully, to sue *The Times* of London and other newspapers for their egregious comments.

But perhaps the strongest evidence for crowding out in pure science is provided by Figure 17.1, which shows a direct correlation between the quality of a nation's science and its GDP per capita. Similar figures show a similar correlation between the quantity of a nation's science (as judged by the numbers of papers it publishes) and its GDP per capita. Figure 17.1 is truly remarkable because it is as pure an example of economic determinism as Karl Marx could have wished: national income determines national expenditure on science. So, though many of the decisions on science funding appear to be taken by politicians (some 85 per cent of academic science within OECD countries is funded by government), the intensity of the economic determinism indicates that, actually, the need is fiercely calibrated by the market, which lobbies government appropriately.

And when, as in Japan, the government supports less than 50 per cent of academic science, the impact as judged by Figure 17.1 is zero, because the market moves swiftly to meet its needs itself. Such tight economic determinism in the face of different governments' policies can be explained only by *tight* crowding out in pure science. These studies thus confirm that science – and civil R&D – are not public goods but, rather, that they behave as private ones.

WOULD IT MATTER IF GOVERNMENTS DID NOT FUND SCIENCE?

We now know that if governments did not fund science, the total quantity and quality of research would be unchanged, because the amount funded by industry and the foundations would increase to fill the gap. Since the domination of any one source of funding is bad for science, the eclipse of government money would be good. That is because science is a collective activity, so scientists tend to be over-orthodox in their thinking.

Under peer review, only those researchers whose ideas are accepted by others will get grants and get published: the heterodox get weeded out, and only by chance do they prosper. Einstein's undergraduate career, for example, was not conventionally successful, so he was sidelined as a patents officer. But Einstein was lucky because, as a theoretician, he needed little more than pencils and paper to flourish. Indeed, some entire disciplines, being theoretical, can be moulded by unorthodox amateurs. Consider linguistics. That owes its very foundation to an amateur, Sir William Jones (1746–94), who was working as a colonial judge in Calcutta when he noted that Sanskrit (the ancient language of the subcontinent), Greek and Latin shared so many words that they must have sprung out of a common root. Another linguistic amateur was Michael Ventris, a British architect, who during the early 1950s revolutionized classical scholarship by showing that the so-called Minoan Linear B script was, in fact, Greek. Ventris was a romantic figure – the hobbyist who solved one of the outstanding problems of ancient archaeology.

To this day, many important scientific debates are initiated by people working outside the universities. Consider social science. Charles Murray, for example, whose 1984 book *Losing Ground* prompted the workfare revolution, is no academic; instead he has worked for a succession of libertarian think tanks. Or consider the environmental movement, which was inaugurated in 1962 when Rachel Carson wrote *Silent Spring*. Yet Carson was no academic; after service with the US Fish and Wildlife Service she was working as a freelance journalist. Or consider Ronald Bailey, whose 1995 book *The True State of the Planet* anticipated Bjorn Lomborg's 2001 book *The Skeptical Environmentalist* by suggesting that economic growth protects the environment because richer societies produce fewer children and can afford to pollute less than do poor societies. His case is strong, if unorthodox, yet Bailey is no academic; he works for the US free market Reason Foundation.

Einstein, Lomborg, Carson *et al.* were successful heretics, and, like most successful heretics, they were theoreticians and thus low maintenance. But experimental science is expensive, so if we are to promote diversity in the lab we have to create a diversity of funding institutions, because such a diversity, by promoting a diversity of grant-reviewing bodies, will help protect the heterodox. We once had diversity. Barbara McClintock, for example, is remembered for the discovery of gene jumping, for which she won the Nobel prize in 1983. But her employer, the Carnegie Institution in Washington , D.C., played its own heroic role, because from 1942 it funded her without imposing the usual instruments of accountability. A maverick, she was spared the conformist peer pressures of grant-writing or paper-publishing: she simply did the experiments she wanted and wrote an annual report.

It is remarkable how many of the recent paradigm-shifters have been mavericks, funded independently. As Carole Jahme reported in her 2000 book *Beauty and the Beasts: Woman, Ape and Evolution*, our understanding of primate intelligence has been transformed by three remarkable women – Jane Goodall (who studied chimps), Dianne Fossey (gorillas) and Birute Galdikas (orang-utans). These women, often known as Louis Leakey's 'trimates' (because Leakey was the womanising primatologist who mentored them) showed that humans were not the only tool-users on the planet. Jane Goodall led the way by showing that chimps used tools, yet she had been merely a secretary – with no academic qualifications – when in 1960 Leighton Wilkie of Des Plaines, Illinois, offered to fund her studies in Africa. But Leighton Wilkie owned a tool-making company, and he *wanted* to believe that tool-making was deeply rooted in our evolutionary past.

No conventional peer review grant body would have funded a young unqualified secretary to disprove what all leading scientists then 'knew' to be true about primates and tool-making. But because Wilkie was rich (and generous – he also funded Fossey and Galdikas) the research was performed to verify his hunch, and our knowledge of our species deepened importantly. But now that the foundations and the government agencies so resemble each other, there are few contemporary McClintocks or Goodalls.

Some enlightened employers survive. In 1998 Novartis signed a five-year contract with Berkeley by which, for $25 million, the company got the first option to commercialize any research coming out of the university's departments of plant and microbial biology. The contract was enlightened, with *The Chronicle of Higher Education* finding that for the academics: 'The deal creates a pool of money for basic, unconventional research of their own choosing, in a way that no grant from a government, a company or a even a foundation ever would'.[19] We need more funders like Novartis, but they will emerge only if government withdraws.

Sometimes the threat to scientific and academic freedom from government is obvious. The McCarthyite era in the US, when academics were forced on pain of losing their jobs to aver loyalty oaths to the federal government, was shocking. State governments could be as frightful. On 30 September 1962 the governor of the state of Mississippi, Ross Barnett, stood on the steps of his state's university, 'Ole Miss', to prevent the admission of James Meredith, its first black student. The federal government was, reluctantly, the redeemer on that occasion, sending 750 federal marshals and 23,000 soldiers to escort James Meredith; in the ensuing riots two people were killed and dozens were injured.

But the real threat from dependence on any dominant source of money is self-censorship. The perennial tragedy of scientists is that, being advocates, they have too often prioritized self-interest over truth. Consider their repeated claims of decline.

THE DECLINE OF SCIENCE, SO-CALLED

Claims of scientific decline are perennial (witness the British Association during the nineteenth century), but they were especially loud during the 1980s. In 1985, for example, Oxford University withdrew the honorary degree it had offered Margaret Thatcher because, the dons claimed, Mrs Thatcher's cuts in university budgets had shrunk British science. Similar claims were simultaneously being made in the US of American scientific decline under Ronald Reagan.

But as I showed at the time, such claims were based on errors.[20] First, they were based on ignorance of crowding out. When government retrenched its funding, the university scientists simply did not 'see' the compensatory increases in private money from the foundations and industry. I personally found it extraordinary to meet the advocates of 'decline' at Oxford, Sussex and other universities and to observe their astonishment when I walked them down the corridors of their own expanding science departments. These people had literally not seen the expansion within their own institutions.

Moreover, it is true that UK and US (and French and German) science is in decline – but only *relatively*. Two centuries ago only two countries did much science, Britain and France. Each, therefore, did about a half of the world's science. Then they were joined by Germany, so for a time each did about a third of the world's science. Then they were joined by the US, so each did about a quarter of the world's science. But now that so many of the countries of the Pacific and Mediterranean rims do science, the UK's and US's share of world science continues to decline. It has, therefore, been easy for the science activists of the UK and US to portray themselves as being in decline.

Yet British and American science continue to expand *absolutely* and hugely; every fifteen or so years the number of their scientists doubles, as do the number of published papers. But the science of the Mediterranean and Pacific rims – reflecting their fast catch-up rates of economic growth – has expanded even faster, with a doubling time of only some ten years. Thus we see that the share of British and American papers published in lead journals is falling but, because science itself grows so fast, the growth of the number of lead journals outstrips the fall in the shares of British and American papers within them. Thus the total number of British and American lead papers continues to grow.

And the share of British and American papers *should* fall. Britain, for example, possesses only 1 per cent of the world's population yet it does 8 per cent of the world's science: in an equitable world it should do only 1 per cent. Similar arguments apply to America's, France's and Germany's science.

The response to my demolition of 'decline' was hostile, because the scientists feared for their government grants. The scientists were using claims of 'decline' to defend their grants, so if there was no 'decline' their lobbying might be weakened. But once it was confirmed that my statistics were indeed correct and that British science was indeed second only to the US in terms of numbers of papers and citations and that the scientists could no longer claim 'decline', the activists simply inverted their arguments. Led by Professor Robert May, the government's Chief Scientist, the science establishment proclaimed that, er, actually, British research was indeed fabulous, so why (here comes the clever bit) shouldn't the government back a winner and invest further in it?[21]

Wasn't that clever? Whether British science is the pits or the star, the activists can always justify more government money. But the intellectual flexibility of scientists in defence of their funding always impresses: consider the Human Genome Project (HGP). That was launched during the late 1980s by the federal agencies in the US and by the Wellcome Trust in the UK. The Human Genome Project proceeded at a deliberate pace until one of its scientists, an impatient man called Craig Venter, left to create his own company, Celera (from the Latin for speed, as in accelerate). Which it certainly did. The HGP was sequencing the genome in a step-by-step fashion, but Venter had recognized the power of computers: he smashed up the genome into lots of different pieces, sequenced them at random, and then used his computers to place the overlapping pieces in the correct order. Soon he was leaving the public scientists behind.

Were they pleased? Did they say 'Hurrah, good old Craig's saving the taxpayers lots of money'? Or did they go ape sh*t, screaming that Craig was a latter-day Blofeld bent on world genome domination? Yup, you guessed it. So they inverted the standard argument: the case for the public funding of the genome project had been that the private sector would never have funded anything as esoteric; when the private sector moved in, the argument switched to claiming that science should be performed by the public sector because nothing could be more dangerous than that the private sector should patent the human genome.

The HGP's argument was economically naive because, in the pharmaceutical industry, no company can invest without IPR. If the HGP's publications precluded companies from patenting DNA sequences, then no company would develop the HGP's research into useful technology, and the whole HGP exercise would have been wasted. Fortunately, the corporations soon found their patents at a separate molecular level, and they now 'own' the functional expression of the genome. Yet the disingenuity of the HGP activists was remarkable – they even complained when Craig Venter incorporated

IN WHICH ADAM SMITH IS VINDICATED BY THE OECD

the HGP's findings into his own results – as if the activists didn't understand what 'public' means.

But behind the froth, the science activists had conceded an historic defeat: they had acknowledged crowding out in pure science. The activists wanted public money, so as to crowd out the private sector. Some 400 years after Bacon had said that pure science represented market failure, the activists conceded the opposite that the market was too damned good at funding the pure science they wanted to keep in the public sector.

The activists will nonetheless identify particular areas that they say the market will never fund. If all science were performed solely by companies, the activists say, then the deleterious effects of cigarette smoking or pollution might be ignored. But civil society in the form of the great foundations and charities creates countervailing research institutions that do not require taxpayers' money. Indeed, the creation of foundations is crowded out by government when it creates its own institutions.

Moreover, government institutions may not protect civil society. They may not introduce democratic accountability into science. Consider the case of GM (genetically modified) foods in the UK today. The government now works so closely with companies that it will use its agencies to promote industry's needs. So the Royal Society, for example, though nominally independent, is largely funded by government (and a recent president, Robert May, was an ex-government Chief Scientist) and the Royal Society follows the government in promoting GM foods tirelessly, even though the British public is hostile and even though some of the environmentalists' fears over GM foods are not groundless.

Activists also argue that governments should fund science for cultural reasons, but the great foundations have done a good job on that score, and there is no evidence for government-funded science being culturally more valuable than the private sector's. And let us not be naive about government: an American government that refused in 2006 to fund stem cell research is no uniquely enlightened patron of research. Indeed, by early 2004 so many senior scientists in America had lost patience with George W. Bush's policies that they wrote a series of open letters of startling language in learned journals, complaining that 'the administration has often manipulated the process by which science enters into its decisions' and denouncing Dr Marburger, the president's science adviser, as a 'prostitute'.[22]

Activists further claim that only government will fund certain types of research. If Langley and not the Wright brothers had been the first to fly an airplane, we would now be told that only government would have supported such a project – just as we are now told that only by government money could the internet or the web have been created. That argument is false. Of course, the American Defense Department made a large investment in the Arpanet (the precursor of the internet), but 'distributed computing' was a technology whose time was coming, and Defense was merely anticipating the inevitable. And, of course, CERN made a huge European government investment in particle physics (which yielded, almost incidentally, Tim Berners-Lee's web technology), but the web was not developed because of government investment in CERN. Rather, CERN found itself in the business of 'distributed computing', so it generated, as a side show to its main business of accelerating particles, solutions to its technological problems. But the market would have uncovered those solutions.

Ironically, government helped delay the development of the net and web. The market – in the shape of customers in the financial, oil and automotive industries – was pushing hard for the new IT technologies, but AT&T and the other telecommunications companies were able to resist the new technologies because legislation had providing them with exclusion powers over the networks.

Doubling time of science: a note

For 250 years the amount of money spent on science and R&D has doubled every fifteen or so years. On 19 July 2007 the National Science Foundation published *Changing US Output of Scientific Articles 1988–2003*,which shows that the expenditure on academic science in the US continued to double over those fifteen years, as it did in the other old science countries of the UK, France and Germany.[23] But although the numbers of scientists and the numbers of published papers has also doubled every fifteen or so years over the last two centuries, that long-term pattern is changing: since 1994 the total number of papers published by the old science countries has hardly risen, and the numbers of their academic scientists have risen by only about 20 per cent. We are, therefore, approaching the era of big science within the mature science countries where more data is published per paper and where more financial (as opposed to human) input is required per paper.

UNIVERSITY SCIENCE UNDER LAISSEZ-FAIRE

If governments did not fund science, university science might be smaller because, though industry and the foundations would, as in Britain after 1981, increase their funding of pure research, they might spend more proportionately within their own laboratories. But because foundation science is public and because industrial science soon becomes public, the cultural consequences of their replacing the government's science would be neutral. And more industrial science would yield an economic benefit, because its initial privacy and in-house location would facilitate its commercial exploitation.

Industrial money for universities does, admittedly, introduce problems. In an infamous case during the 1990s Apotex Inc. tried to prevent Dr Nancy Olivieri, a Toronto University paediatrician, from publishing a study it had funded because Dr Olivieri had shown that Apotex's drug was problematic. But Nancy Olivieri was herself partly to blame because she had signed an unprecedented agreement with Apotex that gave the company veto rights over publication. The lesson of that episode, therefore, is not that universities should eschew companies but, rather, that universities should deal with companies on sensible terms. In 99.9 per cent of cases, that's exactly what happens, to mutual benefit.

But let us acknowledge that some government money might protect academic freedom. The universities face pressures from the companies and foundations that fund their research, and though the many companies and foundations can be played off against each other, a small amount of government-funded science might empower the universities even further against private pressures.

Nonetheless, it might be good to see less research being performed within the universities. The prime function of universities is to teach: of course they should do research, if only to fructify the teaching, but research can be performed in many different places, whereas higher education can take place only within the universities. Should the universities fail to teach properly, therefore, no other institution can compensate. Teaching must, therefore, be the universities' overriding concern, but it is too often subordinated to research and too often outsourced to teaching aides. Less money for research would help focus the universities on teaching.

And universities need to be protected against themselves. Ideally, the universities should be independent centres not just of teaching but – to burnish the teaching with a critical and knowledgeable faculty – of scholarship, fostering scholars who speak truth unto power. Teaching and scholarship are cheap, so they can survive on fees and endowments, so the independent university that criticizes the government on the basis of the careful study of learned works is no hopeless dream. But unfortunately, because of research, that ideal has been lost. In 1810, worried about Britain's lead in the Industrial Revolution, Friedrich Humboldt created the University of Berlin as a research university, and that has been the universal university model ever since.

Research is, of course, a no-brainer for university administrators, because it provides easy markers of success. So the Humboldt University website boasts repeatedly of the twenty-nine Nobel laureates who have worked there, but I can find no obvious reference to its commitment to teaching quality. And the research university needs government money – a dependence that inhibits criticism of government. I bet Humboldt University today would, secretly, foresake its twenty-nine Nobel laureates if it could have boasted of having been an early opponent of Hitler and of eugenics, instead of having fostered so many supporters of both. But as Rousseau said in his 1750 *Discours sur les arts et sciences*, tyrants use science as an opiate, to 'stifle in men's breast the love of liberty'. It was stifled in Humboldt, whose own students helped burn the books from the library in 1933.

To this day, universities rarely criticize government, and the few academics who do knock government are men like Noam Chomsky, who work in independent universities like MIT. Yet the prospects for research freedom globally, however, are looking good. As the share of civil R&D supported by OECD governments falls, so technology research is increasingly free of the state. But academic freedom remains vulnerable. University science budgets being relatively small, governments can meet their expansion. In 1955 the US federal government paid for two-thirds of all academic science, as it did in 2000, even though the actual budget had risen from $140 million to $17 billion (current prices). Academic freedom, therefore, will be maintained only by eternal vigilance.

THE DOG THAT DIDN'T BARK

As it becomes increasingly clear that there is no economic justification for the government funding of science, so the academics have invented further justifications for it. The three most famous economists who claim market failure in science are Edwin Mansfield, Paul Romer and Paul David. Since their papers have been influential let me outline their flaws.

Edwin Mansfield

Mansfield says that 'primary' commercial producers will underinvest in research because their profits are stolen by 'secondary' commercial customers. So, for example, when a seed-breeder (the primary producer) develops a new seed, much of the profit from the new seed is extracted by the farmers (the secondary commercial customers) who plant it. Indeed, surveys have shown that secondary commercial customers can benefit about three times more from innovations than do the primary producers.

So what? The farmers collectively may benefit three times more than the innovating seed-breeder, but one breeder may supply a hundred farmers, so each farmer benefits by only 3 per cent of the breeder. And because the farmers share in the risk of innovation – what if the seeds do not take? – the farmers are, in fact, members of the invisible college of pioneering knowledge-makers, and their profits thus incentivise them appropriately. Mansfield need bear the farmers no grudges, and the primary producers need no compensating government subsidies.

Paul Romer

Romer has gone one step further, believing that the consumers themselves cheat the producers. So, for example, thanks to the producers' research, personal computers have become progressively cheaper and better. But Romer claims that the consumers have contributed nothing to that price-lowering, so they have cheated the researchers, who thus deserve subsidies. Poor Bill Gates – quick, let's give him some subsidies.

Romer's argument, of course, like Mansfield's, supposes that someone's gain is someone else's loss, but that assumption holds only under the myths of the perfect market. And consumers, moreover, are part of the invisible college: the consumer who selects an innovation is taking a risk (*viz* Betamax) and is essentially evaluating it. The consumer of a novel product is, therefore, a partner in the testing of an unproven product and is thus deserving of reward.

One study that chronicled the contribution of consumers to product development was led by Chris Voss of the London Business School. Voss has shown that service to consumers is better in America than in Britain because Americans complain when service is poor. But Britons do not. Britons, when unhappy with service, simply shift their custom, quietly. So British companies have to spend more money on consumer service management than do American companies. To quote *The Economist*, 'a complaint is a gift,' and consumers earn their surplus.[24]

Paul David

The third in our list of unhappy economists is Paul David, who argues that markets fail by 'locking' consumers into inferior technology. Betamax was better than VHS (but VHS succeeded commercially), while the standard QWERTY keyboard is less efficient than one whose letters are placed ergonomically. Therefore government subsidies and direction should supplant markets.

But in their 1999 book *Winners, Losers and Microsoft* the American economists Liebowitz and Margolis showed in a series of case studies that it is simply a myth that markets fail by trapping consumers into inferior technologies.[25] Contrary to myth, there's nothing wrong with QWERTY (which is different in France anyway) and contrary to myth Betamax was inferior to VHS.

Indeed, markets promote cooperation, so they are good at generating common high-quality standards. Consider screws: in 1876 engineers internationally had to choose between the old-fashioned English 55^0 screw thread and the novel American 60^0. As the American engineer William Sellers explained in his *Report to the Franklin Institute* of that year, the world's engineers soon agreed, collectively and voluntarily, on the optimal international standard, without government leadership or subsidy. In Sellers's words: 'The government of France has always been in the habit of interfering with the private habits of people but the American concept of government was that . . . liberty in all things innocent is the birthright of the citizen.' Thus do markets develop common standards, and select the truly superior products, appropriately.

* * *

In *Silver Blaze* Sherlock Holmes remarks on the dog that didn't bark. The equivalent in science funding is that no one – no one – has shown that the government funding of science stimulates economic growth. Adam Smith's faith in laissez-faire would appear to be justified

Epilogue

But scientists, who ought to know
Assure us that it must be so.
Oh, let us never, never doubt
What nobody is sure about.

Hillaire Belloc, *More Beasts for Worse Children*, 1897

One definition of a scientific statement is that it can be tested. The statement 'God is good' cannot be tested. It may be true, but it is not a scientific truth. About 400 years ago Bacon said that science was a public good, which therefore required public funding. In this book I have tested that statement and shown it not to be true. I have therefore used that testing to ask: what actually is science?

I opened this book by agreeing with Bacon that Henry the Navigator was one of the greatest scientists in history. But consider Henry – the real Henry, not the fictitious one of Francis Bacon's imaginations – and see what science really is.

In 1424, when Henry sent his first ship to round Cape Bojador, the only geography Henry knew with confidence was that which he learned by his own explorations and that which he was told by seamen he could trust. And their knowledge was largely tacit, the product of years of experience. Science in Henry's day was not a public good because the ancient texts were often misleading and because few people were publishing. Science was then a private good.

Indeed, the Portugal of Henry's day confirmed that. Portugal supplied entrepreneurs such as Fernão Gomes with monopolies over their discoveries and Portugal, moreover, like all the exploring countries, kept its discoveries secret. So, for example, *The Secret Atlas* of the Dutch East India Company (VOC) was for long an iconic mystery, whose contents were shown only to the trusted captains of the Dutch Republic. Science in 1424, therefore, in a state of nature, was a secretive and almost wholly tacit activity, pursued for private benefit.

Nonetheless, Henry also showed that discoveries could be traded: in 1454, for example, a Venetian sailor, Alvise Ca'da Mosto, was forced by a storm to moor at Sagres. After assessing him, Henry told him about his discoveries. Intrigued, Ca'da Mosto asked to join him; and Ca'da Mosto was to prove a great explorer, surveying the Senegal and Gambia rivers and discovering the Cape Verde Islands and Sierra Leone – and reporting his discoveries back to Henry. In that vignette, therefore, we see how the trading of private and tacit goods between experienced and trustworthy individuals grew into the invisible colleges of today.

But sharing knowledge with the wrong people is a mistake: Ibn Majid, the Arab navigator, was foolish to have shared his secrets with Vasco da Gama, the Arabs' enemy as he is said to have done. Over the years, therefore, scientists have learned to trust with discretion, creating invisible colleges for the purpose. And with their success, so the membership of the invisible colleges has expanded over time, and not just in science. Consider commerce. That, too, was once an invisible college, and in Gresham's day only members were allowed onto the trading floor of the Royal Exchange. Today, the exchanges are much more open. Nevertheless, malefactors are still excluded.

Equally, the Royal Society once monopolized science in Britain. Its experiments were restricted to members of the Society and, in the days before public libraries, the Society's publications were shared with only the select few. Today many people can call themselves scientists; yet science is still an invisible college, and malefactors and the unqualified are still excluded from the scientific societies and journals, and non-scientists still cannot understand the published papers.

Science, like commerce, is a collective activity based on trust, and over evolutionary time we have evolved instincts such as guilt and shame and pride to foster trust. Indeed, these emotions can even be visualized by brain scanning techniques such as functional MRI, and they explain how we evolved to make money, since emotions such as guilt and shame represent the internalization of the cooperation and trust that shift an economy from a Nash equilibrium to a Pareto optimum. One day functional MRI will also visualize the brain areas that account for creativity and entrepreneurialism. I do not want to peddle naive phrenology, but I predict that creativity will localize towards the parts of the brain that control sex, while entrepreneurship will cascade out of the predatory as well as the sexual parts.

But the selfish genes ensure that our enlightened emotions are challenged by Manichean ones, and the triumph of one set of emotions over the other is determined by the social conditioning of childhood. That is why, as the cultural needs of a modern economy are increasingly understood, the costs of admission to the invisible colleges are largely paid in the schools, as pupils are educated in trustworthy behaviour. Consequently, we can easily not 'see' those costs of entry, but a comprehensive economics of science and of commerce needs to encompass the costs of a modern education.

And because we are hard-wired for emotions such as fairness, we will happily pay (as Ultrich Mayr's functional MRI scans showed) taxes for apparently philanthropic activities such as science. But let no one believe that such science is of economic benefit, and let us not forget that science in the hands of undemocratic governments can be a nightmare. As Winston Churchill lamented in his speech of 18 July

417

1940, the nations of the Ribbentrop–Molotov pact were driving humanity 'into the abyss of a new Dark Age, made more sinister, and perhaps more protracted, by the lights of perverted science'.

I do not want to criticize Francis Bacon too much, but let us not forget that his linear model does not exist. Consider Henry the Navigator. He certainly progressed science, but Bacon was wrong to assume that Henry employed scientists and then navigated; actually, Henry navigated first which then stimulated advances in knowledge. His ships sailed 'for God and for profit' (as their sails proclaimed), and only on hitting against the limits of his technology did Henry then command new research, employing scientists opportunistically and testing technology empirically.

Nor is induction the activity that Bacon thought it was. Bacon believed that scientists collected data dispassionately from which they induced theories inexorably. If that were true, then government funding might indeed be of value. But consider Bacon's own ideas on the nature of heat. In his *Instauratio Magna* he wrote: 'Heat is an expansive motion restrained, and striving to exert itself in the smaller particles. The expansion is . . . active and somewhat violent.' That hypothesis, which anticipated the modern kinetic theory by two centuries, could not have been formally induced by the observations available to Bacon – it was an imaginative leap. There is a fluidity to creative scientific thought that no philosopher can capture, and individuals move between so-called pure and so-called applied science in the flash of a synapse.

Equally, Bacon did not see the private funding of research. But Bacon's was not a world of free markets. When, for example, Columbus was looking for sponsors, a private Spanish shipowner called Count Medina Celi offered to support him, but Queen Isabella vetoed that. She was not going to sponsor Columbus – and nor was anyone else. But today we live in a world where anyone may sponsor almost any science, as long as they have the money and as long as it's not stem cell research in George W. Bush's domain.

Science is a creative and logical activity that sprang out of our inherited entrepreneurial instincts that, because it is largely tacit and private in its immediate benefits, has spawned a social structure by which individuals share and trade knowledge to mutual benefit. And it is that social aspect of science that titrates its need. There is no such thing as science, there is only a network of scientists who trade information; and one information they trade is need. If there is a need for more knowledge, the network links the profit-seekers with the researchers, and more research is commissioned. The introduction of politicians into the network may crowd out private sources of funding, but the information still gets transmitted.

Science, indeed, is like sex. Just as sex is selfish but is best pursued with at least one other person, so science is instinctively selfish but we best pursue it socially. The idea of science as a matrix of inter-dependent scientists competing yet also cooperating is not necessarily easy to grasp, but science is not so different from the market. Companies compete, but they also cooperate and can slip into cartels. Yet science can be protected from cartels because the unit of scientific enterprise is the individual researcher, so any potential cartel should be under continual individualistic assault.

It is wonderful that science is such an elaborate social construct. If it were just another public good, it would not speak to the human condition, but because science is so dependent on contradictory human interactions, we can respect it one of humanity's greatest achievements.

And on that cheerful note I shall stop.

Background reading, notes and references

PROLOGUE II

Notes

Gomes Eannes de Azurara's *The Chronicle of the Discovery and Conquest of Guinea* was translated into English by C.R. Beazeley and E. Prestage (Hakluyt Society, London, 1896 and 1899).

P. Russell, *Prince Henry' 'The Navigator': A Life* (Yale University Press, New Haven and London, 2000) is a revisionist biography.

Daniel Boorstin, *The Discoverers* (Random House, London, 1983) is a popular introduction to the early European explorers.

The politics and economics of early European exploration are reviewed in G.V. Scammell, *The World Encompassed: The First European Maritime Empires* (University of California, Berkeley, 1981); W.H. McNeill, *The Pursuit of Power* (University of Chicago Press, 1982); S. Subrahmanyam and L. Thomaz in J. Tracy (ed.) *The Political Economy of Merchant Empires* (Cambridge University Press, 1991); and F. Braudel, *The Mediterranean and the Mediterranean World in the Age of Philip II* (1949, English translation by S. Reynolds, HarperCollins, London, 2nd edn., 1992).

Francis Bacon, *The Advancement of Learning* (1605) and *New Atlantis* (1627), ed. A. Johnston (Oxford University Press, reprinted 1974).

R. Tannahill, *Food in History* (Headline, London, 2nd edn., 2002) is a culinary history.

Reference

1. Quoted in E. Belfort Bax, German Society at the Close of the Middle Ages, 1984, in *Appendix A*.

2. P. Bairoch, 'International Industrial Levels from 1750 to 1980' in *Journal of European Economic History*, 1982, 11: 269-333.

CHAPTER 1

Notes

J. Diamond, *Guns, Germs and Steel* (W.W. Norton & Co., New York, 1998) tells the conventional story of the New Stone Age agricultural revolution.

The controversy over humans and the mass extinctions of animals is reviewed in P. Martin and R. Klein (eds.), *Quaternary Extinctions: A Prehistoric Revolution* (University of Arizona Press, 1984).

The recent molecular and language evidence to support dating the human arrival in America to some 13,000 years ago is reviewed by R. Ward in 'Language and Genes in America' in *Human Inheritance: Genes, Language and Evoltuion*), Bryan Sykes (ed.) (Oxford University Press, 1999)

An account of the history of contraception is provided by R. Tannahill in *Sex in History* (Stein & Day, New York, 1980).

References

1. A. Whiten and W. McGrew, 'Is This the First Portrayal of Tool Use by a Chimp?' in *Nature*, 2001, 409:12.

2. R.J. Britten, 'Divergence between Samples of Chimpanzee and Human DNA is 5%, counting in *Proceedings of the National Academy of Sciences USA*, 2002, 99:13633-5.

3. N. Goren-Inbar *et al.*, 'Evidence of Control of Fire at Gesher Benot Ya'aqov, Israel' in *Science*, 2004, 304:725-7. The authors date the discovery of fire to nearly 800,000 years ago; this remains controversially early; the controversy is discussed in the paper.

4. C.S.L. Lai *et al.*, 'A Forkhead-domain Gene is Mutated in a Severe Speech and Language Disorder' in *Nature*, 2001, 413:519–23.

5. C.S. Henshilwood *et al.*, 'Emergence of Modern Human Behaviour: Middle Stone Age Engravings from South Africa', in *Science*, 2002, 295:1278–80.

6. L.L. and E. Cavalli-Sforza, *The Great Human Diasporas: The History of Diversity* (Addison-Wesley, New York, 1995).

7. K. Harper, *Give Me My Father's Body: The Life of Minik the New York Eskimo* (Simon & Schuster, 2001).

8. P.V. Bradford and H. Blume, *Ota Benga: The Pygmy in the Zoo* (St Martin's Press, New York, 1992).

9. G. Hillman *et al.*, 'Plant-food Economy during the Epipalaeolithic Period at Tell Abu Hureyra, Syria' in D.R. Harris and G.C. Hillman (eds.), *Foraging and Farming* (Unwin Hyman, London, 1989).

10. L.R. Bindord, 'Post-Pleistocene Adaptations' in S.R. and L.R. Binford (eds.), *New Perspectives in Archaeology* (Aldine Press, Chicago, 1968).

11. M. Stiner *et al.*, 'Paleolithic Population Growth Pulses Evidenced by Small Animal Exploitation' in *Science*, 1999, 282:190–94.

12. A.D. Barnosky, 'Assessing the Causes of Late Pleistocene Extinctions on the Continents' in *Science*, 2004, 306:70–75.

13. G. Hardin, 'Tragedy of the Commons' in *Science*, 1968, 162:1243–8.

14. L. Binford, *Constructing Frames of Reference* (University of California Press, 2001).

15. R.D. Lee and R.S. Schofield, in R. Floud and D. McCloskey (eds.), *The Economic History of Britain Since 1700* (Cambridge University Press, 1981).

CHAPTER 2

Notes

Adam Smith's two major books *The Wealth of Nations* (1776) and *The Theory of Moral Sentiments* (1759) have been reprinted by a number of publishers and are easily available.

References

1. C. Renfrew and N. Shackleton, 'Neolithic Trade Routes Realigned by Oxygen Isotope Analysis; in *Nature*, 1970, 228:1062–5.

2. Cavalli-Sforza, op. cit.

3. N. Chagnon, *Yanomamo: The Fierce People* (Holt Rinehart & Winston, 3rd edn., 1983).

4. M.J. Raleigh, M.T. McGuire *et al.*, 'Serotonergic Mechanisms Promote Dominance Acquisition in Adult Male Vervet Monkeys', in *Brain Research*, 1991, 559:181–90.

5. C.B. Stanford *et al.* (including Jane Goodall), 'Hunting Decisions in Wild Chimpanzees' in *Behaviour*, 1994, 131:1–18.

6. B. Winterhalder, 'A Marginal Model of Tolerated Theft', in *Ethology and Sociobiology*, 1996, 17:37–53.

7. F. Jacob and J. Monod, 'Genetic Regulatory Mechanisms in the Synthesis of Proteins' in *Journal of Molecular Biology*, 1961, 3:318–56.

CHAPTER 3

Notes

The story of Egyptian science is told in O. Neugebauer, *Exact Sciences in Antiquity* (Brown University Press, Providence Rhode Island, 1957).

The story of Snefru's river cruise is told in Pierre Montet, *Lives of the Pharaohs* (Weidenfeld & Nicolson, London, 1968).

References

1. Quoted in B. Chatwin, *Songlines* (Jonathan Cape, London, 1987).

2. R. Carneiro, 'A Theory of the Origin of the State' in *Science*, 1970, 169:733–8.

3. Leonard Woolley, *Ur of the Chaldees* (Norton, New York, 1929).

4. C. Boehm, 'Egalitarian Society and Reverse Dominance Hierarchy' in *Current Anthropology*, 1993, 37:763–93.

CHAPTER 4

Notes

The story of money is told in Glyn Davies, *History of Money* (University of Wales Press, 3rd edn., 2002).

Animal intelligence is discussed in S. Parker (ed.), *Self-Awareness in Humans and Animals* (Cambridge University Press, 1995) and in S.M. Reader and K.N. Laland (eds.), *Animal Innovation* (Oxford University Press, 2003).

References

1. From *Commentary on the First Book of Euclid's Elements of Geometry*, quoted in M. Clagett, Greek Science in Antiquity, 1995.
2. *Politics* by Aristotle, translated by T.A. Sinclair 1962, revised T.J. Sanders, Penguin 1981.

CHAPTER 5

Notes

Jean Andreau, *Banking and Business in the Roman World* (Cambridge University Press, 1999) provides a commercial and banking history of Rome.

William H. McNeill, *The Pursuit of Power: Technology, Armed Force and Society since AD 1000* (University of Chicago Press, 1982) provides an economic history of China.

References

1. J. Gimpel, *The Medieval Machine* (Pimlico, London, 1992).
2. R.H. Coase, 'The Nature of the Firm' in *Economica*, 1937, 4:386–405; see also H. Demsetz, 'The Theory of the Firm Revisited' in *Journal of Law, Economics and Organisazion*, 1988, 4:141–62.
3. From Aristotle's *Proptrepticus*, forthcoming publication in *Aristotle and Others on the Study of Philosophy*, Cambridge University Press 2008).
3. R.A. Buchanan, *Technology and Social Progress* (Pergamon Press, Oxford, 1965).

CHAPTER 6

Notes

Lynn White, *Medieval Technology and Social Change* (Oxford University Press, 1962) tells the history of Dark Age technology.

George Grantham, 'Contra Ricardo: On the Macroeconomics of Pre-Industrial Societies', in *The European Review of Economic History*, 1999, 3:199–232, explains why the market stimulated agricultural technological development only in the greater Paris region. It's an important but dense read ('unexhausted scale economies are a sign of an imperfectly competitive market structure').

References

1. F.W. Gibbs 'The History of the Manufacture of Soap' in *Annals of Science*, 1939, 4:169–90.

2. W.H. Long, 'The Low Yields of Corn in Medieval England', in *Economic History Review*, 1979, 2nd series, 32:459–69.

3. George Grantham 1999 Contra Ricardo: *On the Macroeconomies of Pre-Industrial Societies*, *The European Review of Economic History* 3: 199–232.

CHAPTER 7

References

1. Nancy M. Owen, 1966, Thomas Wimbledon's Sermon, *Medieval Studies* 28: 178.

2. Cornelius Walford, 1880, Early Laws and Customs in Great Britain regarding Ford, Transactions of the Royal Historical Society 8: 70–162.

3. D. Friedman, 'The Just Price' in J. Eatwell *et al* (eds.), *The New Palgrave: A Dictionary of Economic Theory and Doctrine* (Macmillan, London, 1987).

4. Maurice Dobb, *Studies in the Development of Capitalism* (Routledge & Kegan Paul, London, 1946).

5. Paul Sweezy, *Theory of Capitalist Development* (Monthly Review Press, New York, 1942.

6. P. Einzig, *History of Foreign Exchange* (St Martin's Press, New York, 1970).

7. S.C. Glfillan 1964, *Invention and the Patent System*, Joint Economic Committee, Congress of the United States (Washington D.C. Government Printing Office).

CHAPTER 8

Notes

The story of early monopolistic capitalism in Europe and of how Holland and England escaped from it is told in Douglass North, *Structure and Change in Economic History* (Norton, New York, 1981).

The story of France's salt trade is told in Simon Schama, *Citizens* (Viking, London and Alfred A. Knopf, New York, 1989) and in Mark Kurlansky, *Salt: A World History* (Walker Publishing, New York, 2002 and Penguin Books, Harmondsworth, 2003).

CHAPTER 9

References

1. A. Maddison, *Phases of Capitalist Development* (Oxford University Press, 1982).

2. M.M. Postan, 'Population and Class Relations in Feudal Societies' in T.H. Aston and C.H.E. Philipin (eds.), *The Brenner Debate* (Cambridge University Press, 1985).

3. E.L. Jones in R. Floud and D. McCloskey (eds.), *The Economic History of Britain 1700–1960* (Cambridge University Press, 1981), pp.66–86.

4. K. Hudson, *Patriotism with Profit: British Agricultural Societies in the Eighteenth and Nineteenth Centuries* (Hugh Evelyn, 1972).

5. Jones. op. cit.
6. G. Heukel in R. Floud and D. McCloskey (eds.), *The Economic History of Britain since 1700* (Cambridge University Press, 1981), pp.182–203.
7. Jones, op cit.
8. C. Pray *et al*, 'Private Research and Public Benefit: The Private Seed Industry for Sorghum and Pearl Millet in India' in *Research Policy*, 1991, 20:315–24.
9. R. Weisheit, *Domestic Marijuana: A Neglected Industry* (Greenwood Press, Westport, Conn., 1992).

CHAPTER 10

Notes

The history of the RNLI is found on its website, www.rnli.org.uk/fs2.

G.I. Brown, *Scientist, Soldier, Statesman, Spy: Count Rumford* (Sutton Publishing, Gloucestershire, 1999) is a biography of Count Rumford.

Samuel Smiles's *Life of George Stephenson* (1875) was reprinted by the Folio Society, London 1975).

In *Education and the State* (Liberty Fund, Indianapolis, 3rd edn., 1994) E.G. West showed that the nationalization of education in Britain during the nineteenth century crowded out philanthropic money disproportionately, thus preventing a rise in education budgets.

References

1. A. Macfarlane, *The Origins of English Individualism* (Blackwell, Oxford, 1978).
2. A. Maddison, *Phases of Capitalist Development* (Oxford University Press, 1982).
3. Ibid.
4. S.D. Chapman, *The Cotton Industry in the Industrial Revolution* (Macmillan, 1972).
5. M. Blaug, 'The Concept of Entrepreneurship in the History of Economics' in *Not Only an Economist: Recent Essays* (Edward Elgar,

Cheltenham, and Northampton, Massachusetts, 1999).

6. C. MacLeod, 'Negotiating the Rewards of Invention: The Shopfloor Inventor in Victorian England' in *Business History*, 1999, 41:17–36.

7. J.D. Bernal, *Science in History* (MIT Press, 1971).

8. Quoted in P. Mathias *The First Industrial Nation* (Methuen, 1983).

9. M. Lindgren, *Glory and Failure: The Difference Engines of Johann Muller, Charles Babbage and Georg and Edvard Sheutz* (MIT Press, Massachusetts, 1990).

10. J. Burnett, *Plenty and Want: A Social History of Diet in England from 1815 to the Present Day* (Nelson, 1966).

11. D. Felix, *Marx as Politician* (Southern Illinois University Press, 1983).

12. R.M. Macleod, 'The Support of Victorian Science' in *Minerva*, 1971, 9:197–230.

13. Ibid.

14. W.T. Harbaugh, U. Mayr, D.R. Burghart 'Neural Responses to Taxation and Voluntary Giving Reveal Motives for Charitable Donations' in *Science*, 2007, 316:1622–5.

145. V. Griskevicius *et al.*, 'Blatant Benevolence and Conspicuous Consumption: When Romantic Motives Elicit Strategic Costly Signals' in *Journal of Personality and Social Psychology* (to be published in late 2007).

16. J. Conaway, *The Smithsonian* (Alfred A. Knopf, New York, 1995).

17. P. Kopper, *America's National Gallery of Art* (Harry N. Abrams, New York, 1991).

CHAPTER 12

Notes

Michael Leapman, *The World for a Shilling* (Review, London, 2001) describes the Great Exhibition of 1851.

K. Colquhoun, *A Thing in Disguise: The Visionary Life of Joseph Paxton* (Harper Perennial, London, 2003) is Joseph Paxton's biography.

Geoffrey Owen, *From Empire to Europe* (HarperCollins, London, 1999) compares German and American capitalism during the nineteenth century.

References

1. Michael Leapman, *The World for a Shilling*, Review, London 2001.
2. Lynn Playfair, *Journal of the Society of Arts*, 1867, 15:477.
3. H. Gospel (ed.), *Industrial Training and Technical Innovation: A Comparative and Historical Study* (Routledge, London, 1991).
4. *OECD Historical Statistics 1970–1999* (OECD, 2000 edn.).
5. Michael Porter, Hirotaka Takeuchi and Mariko Sakakibara, *Can Japan Compete?* (Basic Books, New York, 2000).
6. M. Jansen, *The Making of Modern Japan* (Harvard University Press, 2001).
7. R.M. Macleod, 'The Support of Victorian Science' in *Minerva*, (1971, 9:197–230).
8. A. Maddison, *Phases of Capitalist Development* (Oxford University Press, 1982).
9. B.R. Mitchell, *European Historical Statistics* (Cambridge University Press, 1975). J.M. Winter, 'The Decline of Mortality in Britain 1870–1950' in T. Barker and M. Drake (eds.), *Population and Society in Britain 1850–1980* (Batsford, London, 1982). J.M. Winter *The Great War and the British People* (Harvard University Press, 1986).
10. Quoted in J. Parry *The Rise and Fall of Liberal Government in Victorian Britain* (Yale University Press, 1993).
11. D. Noble, 'The Paradox on Statistics on Science and its Funding' in *The Biochemist,* 1989, 11:13–15.
12. H.W. Richardson, 'Chemicals' in D.H. Aldcroft (ed.), *The Development of British Industry and Foreign Competition 1875–1914* (Allen & Unwin, London, 1968).
13. P. Lindert and K. Trace in D. McCloskey (ed.), *Essays on a Mature Economy* (Princeton University, 1971).
14. Quoted in A.J. Ihde, *The Development of Modern Chemistry* (Harper & Row, New York, 1964).

CHAPTER 12

Notes

The story of America's early science policies, including the extracts from the protagonists' private letters, is provided by N. Reingold, 'Science in the Civil War' in *Isis*, 1958, 49:307–18, and by D.J. Kevles, 'The First World War and the Advancement of Science in America' in *Isis*, 1968, 59:427–37.

An introduction to Justin Morrill's mercantilism is provided by T. DiLorenzo, *The Real Lincoln* (Prima Lifestyles, 2002).

The story of America's inter-war science policies is provided in *Science Policy Background No. 1*, (US Government Printing Office, 1986).

The story of America's post-war science policies is provided by H. Dupree, *Science in the Federal Government* (Harvard, 1957) and by J.L. Perrick *et al.*, *The Politics of American Science* (MIT, 1972).

References

1. Quoted by Heather Ewing, *The Lost World of James Smithson: Science, Revolution and the Birth of Smithsonian*, Bloomsbury, 2007.

2. Adam Smith, from a Letter to Dr Cullen, quoted in John Rae, *Life of Adam Smith*, Macmillan, London 1895.

3. D.H. Calhoun, *The American Civil Engineer: Origins and Conflict* (Harvard University Press, 1960).

4. E. Fano, 'Technical Progress as a Destabilizing Factor and as an Agent of Recovery in the US between the Two World Wars' in *History and Technology*, 1987, 3:249–74

5. Quoted in W. Isaacson, *Einstein* (Simon & Schuster, New York, 2007).

6. Fano, op. cit.

7. R.M. Solow, 'Technical Change and the Aggregate Production Function' in *The Review of Economics and Statistics*, 1957. 39:312–20.

8. Quoted in *Science Policy Background Report* No. 1, US Government Printing Office, Washington D.C. 1996.

9. R.D. Lapidus, 'Sputnik and its Repercussions' in *Aerospace Historian*, 1970, 17:89

10. www.nsf.gov/sbe/srs/infbrief/nsf02326/start.htm
11. J.D. Adams, *et al.*, *The Influence of Federal Laboratory R&D on Industrial Research* (NBER Working Paper No.w7612, 2000).
12. N. Ferguson, *Pity of War* (Basic Books, New York, 1999).

CHAPTER 13

Notes

R. Macleod, 'The Support of Victorian Science, *Minerva*, 1971, 9:197–230, provides the story of Britain's Victorian science policies.

C.H. Shinn, *Paying the Piper* (Falmer Press, xxx, 1986) provides the history of British government funding for the universities.

B. Pimlott, *Harold Wilson* (HarperCollins, London, 1992) provides the quotes from politicians fearing Russian economic dominance.

D. Kevles, *In the Name of Eugenics* (Harvard University Press, 1995) provides a history of eugenics.

P. Feyerabend, *Against Method* (Verso, London 1975) is a post-modern account of science.

M. Walker and T. Meade (eds.) *Science, Medicine and Cultural Imperialism* (Macmillan, London, 1991) and G.W. Craig, *The Germans* (Penguin, Harmondsworth, 1991), tell the story of the Nazi Party penetration of science and the universities.

References

1. J. Harwood, 'Genetics, Eugenics and Evolution' in *British Journal of the History of Science*, 1989, 22:257–65.
2. Quoted in D.J. Kelves *In the Name of Eugenics*, Alfred A. Knopf, 1985.
3. L. Thomson, 'Origin of the British Legislative Provision for Medical Research' in *Journal of Social Policy*, 1973, 2:43; L. Bryder, 'Tuberculosis and the MRC' in J. Austober and L. Bryder (eds.), *Historical Perspectives on the Role of the MRC* (Oxford University Press, 1989)
4. J. Austober, 'Walter Morley Fletcher' in J. Austober and L. Bryder op. cit.

5. www.amrc.org.uk
6. M. Weatherall and H. Kamminga, *Dynamic Science: Biochemistry in Cambridge 1898–1949* (Cambridge Wellcome Unit, 1992).
7. Quoted in D.J. Kelves *In the Name of Eugenies*, Alfred A. Knopf, 1985.
8. Quoted by C. Barnett, *Audit of War* (Macmillan, London, 1986).
9. D. Swinbanks, 'Japan Overtakes Soviet Union in Research Spending League' in *Nature,* 1987, 325:188.
10. *Times Higher Education Supplement*, 18 November 1994, p.14.
11. A. Saegusa, 'Japanese Technology Fund Faces Ministry Criticism' in *Nature*, 1997, 386:424.

CHAPTER 14

Notes

D.A. Irwin & P.J. Klenow, 'High Tech R&D Subsidies: Estimating the Effects of Sematech' in *Journal of International Economics*, 1996, 40:323–44, shows that Sematech failed. Professor Flamm of the University of Texas disagreed, yet even he was forced to conclude that: 'the data seem to suggest that Sematech reduced the R&D expenditure somewhat' (K. Flamm and Q. Wang, 'Sematech Revisited' in C.W. Wessner (ed.), *Securing the Future*, National Academy of Sciences, Washington, D.C., 2003; available freely on the web).

R.M. Solow, 'Technical Change and the Aggregate Production Function' in *The Review of Economics and Statistics*, 1957, 39:312–20, shows how technology creates wealth.

References

1. E. Mansfield, 'Academic Research and Industrial Innovation' in *Research Policy*, 1991, 20:1–12.
2. Report to the National Science Foundation, *Indicators of International Trends in Technological Innovation*, 1976.
3. C. Freeman and L Soete, *The Economics of Industrial Innovation* (Pinter, London, 1997).

4. D. Hicks & S. Katz, *The Changing Shape of British Industrial Research* (Sussex University Press, 1997).

5. Institute of Scientific Information in *Current Contents*, 1994, 37:4.

6. E. Mansfield, 'Basic Research and Productivity Increase in Manufacturing' in *American Economic Review*, 1980, 70:863–73.

7. Z. Griliches, 'Productivity, R&D and Basic Research at Firm Level in the 1970s' in *American Economic Review*, 1986, 76:141–54.

8. M. Lynn, *The Billion Dollar Battle: Merck v Glaxo* (William Heinemann, London, 1991).

9. W.M. Cohen and D.A. Levinthal, 'Innovation and Learning: The Two Faces of R&D' in *Economic Journal*, 1989, 99:569–96.

10. See discussion in P. Stephan, 'The Economics of Science' in *The Journal of Economic Literature*, 1996, 34:1199–262.

11. Lynn, op. cit.

12. E. Mansfield *et al.*, 'Imitation Costs and Patents: An Empirical Study' in *Economic Journal*, 1981, 91:907–18.

13. G.J. Stigler, 'The Economics of Information' in *Journal of Political Economy*, 1961, 64:213–25.

14. www.nsf.gov/sbe/srs/stats.htm

15. T. Burnham, 'High-testosterone Men Reject Low Ultimatum Game Offers' in *Proceedings of the Royal Society B: Biological Science*, 2007, 274:2327–30

16. Michael Noad *et al.*, 'Cultural Revolution in Whale Songs' in *Nature*, 2000, 408:537.

17. E. von Hippel, 'The Economics of Product Development by Users' in *Management Science*, 1998, 44:629–44; E. von Hippel, *The Sources of Information* (Oxford Univerity Press, 1988).

18. T. Allen *et al.*, 'Transferring Technology to the Small Manufacturing Firm' in *Research Policy*, 1983, 12:199–211.

19. L. Galambos and J.L. Sturchio, 'The Transformation of the Pharmaceutical Industry in the 20th Century' in J. Krige and D. Pestre (eds.), *Science in the 20th Century* (Harwood Academic Publishers, Paris, 1997).

20. A.L. Saxenian, *Regional Advantage* (Harvard University Press, 1994, p.33).

21. J.J. Wallis and D.C. North, 'Measuring the Transaction Sector in the American Economy, 1870–1970' in S.L. Engerman and R.E. Gallman (eds.), *Long-Term Factors in American Economic Growth* (University of Chicago, 1987).

22. Quoted by R.M. Huber, *The American Idea of Success* (McGraw-Hill, New York, 1971).

23. S. Macaulay, 'Non-Contractual Relations in Business: A Preliminary Study', in *American Sociological Review*, 1963, 28:55–67.

24. J.K. Rilling *et al.*, 'A Neural Basis for Social Cooperation' in *Neuron*, 2002, 35:395–405

25. B.C. Martinson, M.S. Anderson and R. De Vries, 'Scientists Behaving Badly' in *Nature*, 2005, 435:737–38.

26. Stephan op. cit.

27. D. de S. Price, *Little Science, Big Science* (Columbia University Press, New York, 1963).

CHAPTER 15

References

1. I. Busom, 'An Empirical Evaluation of the Effects of R&D Subsidies' in *Economics of Innovation and New Technology*, 2000, 9:111–48.

2. K.L. Sokoloff, 'Inventive Activity in Early Industrial America' in *Journal of Economic History*, 1988, 48:813–50.

3. R. Blundell *et al.*, 'Market Share, Market Value and Innovation in a Panel of British Manufacturing Firms' in *Review of Economic Studies*, 1999, 66:529–54.

4. F. Etro, 'Innovation by Leaders' in *Economic Journal*, 2004, 114:281–303.

5. S.J. Gould, *The Panda's Thumb* (W.W. Norton & Co., New York, 1980), pp.183.

6. P. Romer, 'Endogenous Technical Change' in *Journal of Political Economy*, 1990, 98:S71–S102.

7. Quoted in H.A. Meier, 'Thomas Jefferson and a Democratic

Technology' in C.W. Pursell (ed.) *Technology in America* (MIT Press, 1986).

CHAPTER 16

References

1. *The Economist*, 22 October 2005.
2. W. Baumol, *Entrepreneurship, Management and the Structure of Payoffs* (MIT Press, 1993).
3. E. Mansfield, 'Academic Research and Industrial Innovation', in *Research Policy*, 1991, 20:1–12.
4. T. Astebro, 'The Return to Independent Invention: Evidence of Unrealistic Optimism, Risk Seeking or Skewness Loving?' in *Economic Journal*, 2003, 113:226–39
5. P. Grootendorst and L. Di Matteo, 'The Effect of Pharmaceutical Patent Term Length on Research and Development and Drug Expenditures in Canada' in *Healthcare Policy/Politiques de Santé*, 2007, 2:63–84.

CHAPTER 18

Notes

To determine the effects of the different national business:government ratios on the funding of national civil R&D, a multiple linear regression analysis using a forward selection procedure was performed. Two unadjusted Pearson correlation coefficients between percentage of GDP spent on civil R&D and GDP per capita and the ratio of business:government funding for civil R&D were $r = 0.69$, t-value 2.469 ($p = 0.024$) and $r = 0.80$, t-value 4.103 ($p<0.001$). 85 per cent of the variation in the amount of GDP spent on civil R&D was explained by the GDP per capita and the ratio of business:government funding for civil R&D. This suggests crowding out for civil R&D budgets, with $1 of public funding crowding out more than $1 of private funding.

References

1. P. David and B. Hall, 'Heart of Darkness: Public–Private Interactions Inside the R&D Black Box' in *Research Policy*, 2002, 29:1165–83.

2. www.nsf.gov/sbe/srs/stats.htm

3. Advisory Council on Science and Technology (ACOST), *Developments in Biotechnology* (HMSO, London, 1990).

4. T. Kealey, *The Economic Laws of Scientific Research* (Macmillan, London, 1996).

5. D. da S. Price, *Little Science, Big Science* (Columbia University Press, New York, 1963).

6. www.nsf.gov/statistics/seind00/access/c2/c2h.htm

7. T. Kealey, 'The Economic Laws of Research, in *Science and Technology Policy*, 1994, 7:21–7; T. Kealey, *The Economic Laws of Scientific Research* (Macmillan, London, 1996).

8. W. Park, 'International R&D Spillovers and OECD Economic Growth' in *Economic Enquiry*, 1995, 33: 571–90.

9. P.A. David, 'From Market Magic to Calypso Science Policy: A Review of Terence Kealey's "The Economic Laws of Scientific Research" in *Research Policy*, 1997, 26:229–55.

10. T. Kealey, 'Why Science is Endogenous: A Debate with Paul David' in *Research Policy*, 1998, 26:897–923.

11. T. Kealey, 'Cause and Defect' in *Nature*, 1996, 383:474.

12. L.K. Cheok, 'Pump Priming Works in Singapore' in *Nature*, 1996, 384:508.

13. OECD report, quoted in *Times Higher Education Supplement*, 22 November 1991, p.1.

14. E. Fano, 'Technical Progress as a Destabilizing Factor and as an Agent of Recovery in the US between the Two World Wars' in *History and Technology*, 1987, 3:249–74.

15. See notes to Chapter 15.

16. Quoted in *Nature*, 1993, 364:742.

17. F. Narin and D. Olivastro, 'Status Report: Linkage Between Technology and Science' in *Research Policy*, 1992, 21: 237–49.

18. R. Agarwal and M. Gort, 'First Mover Advantage and the Speed of Competitive Entry' in *Journal of Law and Economics*, 2001, 44:161–77.

19. G. Blumenstyk, 'A Vilified Corporate Research Partnership Produces Little Change Except Better Facilities' in *The Chronicle of Higher Education*, 22 June 2001, p.A24.

20. T. Kealey, 'Government-Funded Academic Science is a Consumer Good' in *Scientometrics*, 1991, 20:369–94.

21. R. May, 'The Scientific Wealth of Nations' in *Science*, 1997, 275:793–6.

22. Anon., *The Economist*, 10 April 2004, pp.75–6.

23. http://www.nsf.gov/statistics/nsf07320/

24. Anon., 'A Complaint is a Gift' in *The Economist*, 24 April 2004, p.73.

25. S.J. Liebowitz and S.E. Margolis, *Winners, Losers and Microsoft: Competition and Antitrust in High Technology* (Independent Institute, Oakland, California, 1999).

Legends to the graphs

Figure 11.1 and Figure 11.2

The data came from Angus Maddison 2003 The World Economy: Historical Statistics OECD. The GDP per capita is in 1990 international Geary-Khamis dollars.

Figure 12.1

The federal expenditure on basic science, with the exception of that for the year 1940, comes from D.C. Mowery and N. Rosenberg (1989) *Technology and the Pursuit of Economic Growth* (Cambridge: Cambridge University Press); 1940 data came from *Federal Funds For Research, Development and other Scientific Activities* 1972 National Science Foundation, Washington DC. The GDP *per capita* in 1990 international Geary-Khamis dollars comes from Angus Maddison 2003 The World Economy: Historical Statistics OECD

Figure 17.1

The relative citation data, which reflects the quality of science, comes from T Braun, W Glanzel & A Schubert 1987 One More Version of the Facts and Figures on Publication Output and Relative Citation Output of 107 Countries 1978–1980 *Scientometrics* 11:9–15, and the GDP *per capita* data from *OECD Economic Surveys, UK* 1981 OECD Paris. Graphs comparing the quantity of national science with GDP per capita look very similar (T Kealey 1996 *The Economic Laws of Scientific Research* Macmillan 1996).

Key: A = Austria; Al = Australia; B = Belgium; C = Canada; CH = Switzerland; D = Denmark; E = Spain; Fi = Finland; F = France; G = Germany; Gc = Greece; I = Italy; Ir = Ireland; J = Japan; N = Norway; Ne = Netherlands; NZ = New Zealand; P = Portugal; S = Sweden; T = Turkey; UK = Great Britain; US = USA.

Acknowledgments

First I thank my wife Sally and our children, Helena and Teddy, who have tolerated my filling every empty moment with this book. No man could be luckier in his family.

I thank Matt Ridley for being a champion when I needed support. I also thank Professor Paul Luzio of the University of Cambridge, Dr Chris Keightley of Cambridge Bioclinical Ltd, Professor David Edgerton of Imperial College, and Professor Martin Ricketts, Sir Martin Jacomb, Professor John Clarke and Mrs Colleen Carter of the University of Buckingham for being friends in need.

I thank Professor Fred Miller and Professors Ellen and Jeffrey Paul of the Social Policy and Philosophy Centre at Bowling Green State University, Ohio, who provided me with a key sabbatical that allowed me to start this book. Further, I thank Fred Miller for introducing me to the philosophy of science of ancient Greece, and Pierre Desrochers of the University of Toronto for educating me in scientific co-operation between companies.

Finally, I thank the staff at William Heinemann, who have been wonderful. Ravi Mirchandani and Caroline Knight have been brilliant editors, Lydia Derbyshire, a brilliant copy editor, and Alban Miles knows how to get authors actually to submit manuscripts.

I have been an infuriating author as, distracted by a proper job, I have missed deadline after deadline, so I am dedicating this book to Mrs Felicity Bryan, my agent, whose tolerance of my faults has been superhuman.

Index

Britain
  budget 337
  chemical industry 224–6
  economic growth 211–13
  electronics industry 225–6, 291
  empire 203
  and eugenics 266
  iron and steel industry 220–4
  market economy 153
  parliamentary rule 142–4
  population 151, 152, 166–7
  poverty 151–3
  rebellious nature of people 164–5,
    166
  science/research 257–88, 393, 404–5
  Stuart reign 142, 144
  taxes 145
  technological decline 206–10
  Tudor reign 141–2
  university funding 278–80, 281–4
British Association (for the
    Advancement of Science) 258
Broghill, Lord 329
Bronze Age 63–73, 82, 316
Brunel, Isambard Kingdom 170
Bruno, Giordano 20
Buffett, Warren 197
Bury, J.B., *The Idea of Progress* 3
Bury St Edmund's, Abbot of 138
Bush, Vannevar 244–8, 275
business schools 255
buttons 134

Ca'da Mosto, Alvise 416
calcium-channel antagonists 272–3
Calder Hall 285
caloric theory 180, 191
Cambridge University 48–9, 180, 190,
    197, 276, 279
Canaanites 75, 77–8
Canada 213

cancer 251
Cancer Research Campaign 276
Cape Bojador 9–10, 25
Cape of Good Hope 11
capital punishment 239–40
capitalism
  early monopolistic 138–40
  and economic growth 210–14
  Rhenish 218–19
carbon dioxide 175
Cardana, Girolama 134, 135–6
Carnegie, Andrew 196, 233, 254
Carnot, Sadi 180, 295
Carson, Rachel 401
cartel model 218–19, 292
Cartwright, Rev. Edmund 169
caste systems 68–70
Cavendish, Henry 189
Cavendish, William, Duke of
    Devonshire 260
Cavendish Laboratories 260
Celera 405
censorship 108–9
Ceuta 7–9
Chamberlen, Peter 307
Charlemagne 122, 123–4
Charles II 329
Chatsworth 204
Chaucer, Geoffrey 126
chemical industry 224–6
Cheops, Pharaoh 61
cheques 131
Chesterton, G.K. 266
Childe, V. Gordon 34, 38
chimpanzee 56–7, 82
China 103–6, 110
  agriculture 117
  early civilization 64–5
  manufacturing output 26
  technology 36, 103–4
Chomsky, Noam 411

chronometer 257
Church
  eastern base of 24
  encouragement of markets 122
  and eugenics 266
  regulation of printing 108
  and science 20–1, 260
Cistercian monks 123
Citizens Committee for the Conquest
  of Cancer 251
Clay, Henry 237
Cleopatra 99
climate change and the New Stone Age
  38
club good 335–6
clustering 311–14
Cobbett, William, *Rural Rides* 186
coins 75–6, 78–9, 115
Colbert, Jean Baptiste 181–2
Cold Spring Harbor Laboratory 265
Cold War 247, 248
Cole, Sir Henry 'King' 203–5
Coleridge, Samuel Taylor 261
Columbus, Christopher 11–12
command economies 76
companies
  clustering 311–14
  co-operation 366–8
  funding of science 159–60, 296–9,
    303–4, 306
  information-sharing 313–14
  problems of specialization 96–9
competition 345–53
computers 285, 291–3
conferences 311, 314, 315, 326
confidentiality agreements 376
Constantine, Emperor 24
Constantinople 24
consumers, diminishing returns 339
continental drift 332
contraception 45

convergence 214, 218, 219, 229, 230,
  389
cooperation 319–20
Copernicus, Nicolaus 269
copper 63
copying 87–8, 300–4, 305–6, 388,
  396
cotton industry 168–9, 232–3
Courtauld, Samuel 197
cowrie shells 76–7
crank 112
creative destruction 351, 353
Cro-Magnon man 33
Croesus 78
Crompton, Samuel 169
crop rotation 116, 117–18
crowding in/crowding out 380–400
Crystal Palace 204
Ctesibius 99, 101
CUDOS 334
currency 76–9, 104, 105, 131
Curtiss, Glenn 363
cystic fibrosis 267–8

da Gama, Vasco 6, 12–14, 416
Dahomey 65
dam construction 63
Dark Ages 111–20
Darwin, Charles 54, 71, 185, 189, 258,
  259, 307–8, 361
Davenport, Charles 265, 266
David, Paul 413–14
Davy, Humphrey 192–3
Davy–Faraday Laboratory 263
de Caus, Salomon 173–4
deduction 80
definitions 80
Del Ferro, Scipione 134, 135
democracy 85–6
Democritus 3
Denmark 271–2

metallurgy 75
Michell, Rev. Dr John 49
Microsoft 292, 351–2, 364, 375, 386
Midas 78
Middlesex Canal 233
Miletus 87–9
Mill, John Stuart, *Principles of Political Economy* 167
millennial celebrations 205–6
Millennium Dome 205–6
Ming dynasty 103, 105–6
Minoan civilization 64
Mohenjo-Daro 64
Mond, Alfred 263
money-lending 126–7, 131
monopolies 342
  battle against 147–8
  and competition 345–9
  early monopolistic capitalism 138–40
  and niches 349–53
Moray, Sir Robert 329
Morton, Cardinal John 141–2
Morton's fork 141–2
movable type 103, 108
MRC 274–8, 399
Murray, Charles 401
Mushet, David 194, 202
Mushet, Robert 194, 202
Muslims
  encounter with Vasco da Gama 13
  philanthropy 198
  response to printing 108, 109
  and the sack of Ceuta 8
  scientific advances 106–7
  scientific decline 107–8
  spice trade 5–7

NASA 248, 249
National Academy of Sciences 234–5, 237, 240

National Aeronautics and Space Administration 248, 249
National Defense Research Committee 245
National Institutes of Health 246, 251
National Public Radio 199–200
National Research Council 240–1
National Science Foundation 246–8, 250
national wealth 167–8
Natural and Political Observations and Conclusions upon the State and Condition of England, 1696 151–3
natural selection 259, 307, 360–1
Nature 259
NDRC 245
Neanderthals 33
Netherlands 140–1, 144, 145, 146–7, 153, 266, 371
New Stone Age Revolution 34–40, 43, 46–7, 51–3, 70, 117
New Zealand 392
Newcomen, Thomas 172–3, 175, 183
Newton, Isaac 190
niches 349–53
nitrocellulose 227–8
nitrogen fixation 116, 117
Nobel prizes 295, 296
Norfolk four-course system 118
Novartis 402
NPR 199–200
NRC 240–1
nuclear power 285

observation 80
obstetric forceps 307
OECD 18–20, 391, 392–3, 394
Office of Scientific Research and Development 245, 246
oligopolies 342
Olivieri, Nancy 409